SpringerBriefs in History of Science and Technology

More information about this series at http://www.springer.com/series/10085

Helge Kragh

From Transuranic
to Superheavy Elements

A Story of Dispute and Creation

 Springer

Helge Kragh
Niels Bohr Archive
Niels Bohr Institute
Copenhagen
Denmark

ISSN 2211-4564   ISSN 2211-4572   (electronic)
SpringerBriefs in History of Science and Technology
ISBN 978-3-319-75812-1   ISBN 978-3-319-75813-8   (eBook)
https://doi.org/10.1007/978-3-319-75813-8

Library of Congress Control Number: 2018932168

Printed on acid-free paper

This Springer imprint is published by the registered company Springer
International Publishing AG part of Springer Nature
The registered company address is: Gewerbestrasse 11, 6330 Cham, Switzerland

Preface

This essay is about the endeavours of nuclear scientists to produce and investigate chemical elements beyond uranium and especially those "superheavy" elements that lie at the end of the periodic table. My interest in the subject was aroused when I studied Niels Bohr's semi-classical atomic theory and his application of it in the early 1920s to the electron configuration of heavy elements. Bohr even suggested an atomic model for the then dreamlike hypothetical element with atomic number 118—the same element which received official recognition in 2016 and is today the heaviest of all known elements. The amount of scientific literature on superheavy elements and related topics is frighteningly large, but by far most of it consists of technical papers whereas only little has been written from a historical or general perspective. In fact, historians and philosophers of science have shown almost no interest in the subject at all, something I hope the present work will help rectifying.

I would like to emphasise that the work is in no way intended to be a full history of how the science of superheavy elements has evolved. A history it is, but a fragmented one. Not only do I not deal with all the transuranic elements individually and in detail, there are also large parts of the development which I leave out or only mention cursorily. Thus, I ignore the technical apparatus—such as accelerators and detector systems—which have been of overwhelming importance in the synthesis of the very heavy elements since the 1940s. Detection methods based on nuclear chemistry are another and no less important part of the development that I have left out.

My focus is not so much on the scientific and technical aspects of this interdisciplinary branch of research as it is on conceptual and contextual aspects, including the criteria of element discovery and the inevitable priority controversies that these criteria are supposed to make an end to. In the last chapter, which relies on a recent paper in the journal *Substantia*, I argue that the history of heavy synthetic elements provides a valuable case study from the point of view of philosophy and sociology of science. For example, it problematises the notion of discovery and also the very concept of a chemical element. Moreover, the history is instructive with regard to the sometimes strained relationship between chemistry and physics. *From Transuranic to Superheavy Elements* is a story of how the periodic table was successfully expanded with new elements, but it is also a story of many mistakes and failed discovery claims. The latter category is no less interesting than the first, and I deal with it in some detail. For comments on and help with the manuscript, I want to thank Dieter Hoffmann and Sabine Lehr.

Copenhagen, Denmark   Helge Kragh
November 2017

Contents

Chapter 1
Beyond Uranium, Ca. 1890–1950

Abstract The transuranic elements at the end of the periodic table have since the 1960s been known as superheavy elements. Until 1939 no element heavier than uranium was known and yet there was in the earlier period considerable interest in the possible existence of such elements. The interest was in part of a speculative nature and in part based on calculations of the electron structure of heavy elements. After all, why should uranium be the heaviest element in nature? Although transuranic elements were "known" theoretically in the 1920s, it was only with the development of nuclear physics and technology in the following decade that the first of these elements—neptunium and plutonium—were actually produced. By 1951 six transuranic elements had been added to the periodic table, all of them by an innovative group of Californian nuclear scientists.

Keywords Transuranic elements · Periodic table · Niels Bohr
Enrico Fermi · Glenn Seaborg · Neptunium · Plutonium

Ideas of chemical elements even heavier than uranium go back to the late nineteenth century, where they emerged in connection with speculations concerning the end of the periodic system. Following the first quantum-based theories of atomic structure several physicists suggested electron configurations of hypothetical transuranic elements. But although there were theories of such elements by 1930, it was only with Fermi's pioneering neutron experiments a few years later that it seemed realistic to produce synthetic transuranic elements. The first man-made elements date from the late 1930s and by 1950 the periodic table had been extended with several elements beyond uranium. Although the term "superheavy elements" had not yet been coined, the development towards artificially produced and still heavier elements had begun.

H. Kragh, *From Transuranic to Superheavy Elements*, SpringerBriefs in History of Science and Technology, https://doi.org/10.1007/978-3-319-75813-8_1

1.1   Introduction

The discovery of the element uranium is often credited to the German chemist Martin Heinrich Klaproth who in 1789 recognised that the mineral pitchblende contained an unknown metal. Referring to William Herschel's sensational discovery of the planet Uranus a few years earlier, he proposed to the call the new metal uranium (Weeks and Leicester 1968, pp. 266–271). However, Klaproth's attempts to isolate the metal failed and it took until 1841 before Eugène-Melchior Peligot in France could produce a powder of pure uranium. In Mendeleev's original periodic table of 1869 the atomic weight of the metal was given as 116, but in a slightly later version he more than doubled the value (Scerri 2007, p. 129). He cited it as $M = 240$, thus making uranium the heaviest of all known elements. Over the next couple of decades it was generally accepted that no other element has an atomic weight as large as or larger than uranium's. The metal discovered by Klaproth (or perhaps by Peligot) marked the end of the periodic table. And this was not the only peculiarity exhibited by uranium, for in early 1896 Henri Becquerel demonstrated that uranium and its salts give off a new kind of rays. Radioactivity was discovered.

With atomic number $Z = 92$ and atomic weight $M = 238.0289$, uranium is the heaviest of the naturally occurring chemical elements listed in the periodic system. Strictly speaking, also the transuranic elements neptunium and plutonium occur naturally. However, the trace amounts found of these two elements are not of primordial origin but owe their existence to nuclear reactions in uranium or thorium such as neutron capture followed by beta decay. The two elements exist in nature in extremely low concentrations only, such as illustrated by the amount of plutonium in the uranium minerals pitchblende and monazite, which is about one part to 10^{11}. For neptunium the ratio is even smaller, about one part to 10^{12}. Some elements of lower atomic number, especially technetium (Tc, $Z = 43$) and promethium (Pm, $Z = 61$) but also astatine (At, $Z = 85$), francium (Fr, $Z = 87$) and protactinium (Pa, $Z = 91$), are also practically non-existent in terrestrial nature (Scerri 2013).

Beyond uranium the periodic table includes presently no less than 26 elements which have all been manufactured in the laboratory and the best known of which is plutonium of atomic number 94. The heaviest of the transuranic elements are often called "superheavy elements," a term with no precise meaning but which often refers to the transactinide elements with Z ranging from 104 to 120. In some cases the term is used only for $Z > 110$ and in other cases it just refers to very heavy elements. What these elements have in common, as seen from a modern perspective, is that they owe their existence to effects due to the shell structure of the atomic nucleus. So far the last of the known superheavy elements is $Z = 118$, a substance which received official recognition as an element in 2016 and is named oganesson, chemical symbol Og.

The term superheavy element (SHE or SHEs in plural) largely owes its origin to the American physicist John Wheeler, who in the 1950s examined theoretically the limits of nuclear stability. However, the term can be found even earlier, first in an informative review paper written by the American chemist Laurence Quill (1938) at a time when uranium was still the heaviest known element. Quill's reference to "super-heavy atoms" was to possible transuranic elements with atomic number up to 96. Two years later *Science News Letter* announced prematurely that "super-heavy element 94 [is] discovered in new research" (Anon 1940). More interestingly, in 1942 George Gamow speculated that the original nuclear matter of the universe would undergo fission and produce what he called unstable superheavy nuclei (Gamow 1942).

On the basis of simple extrapolations from the liquid drop model of fission, Wheeler (1955, p. 181) suggested that atomic nuclei twice as heavy as the known nuclei might be ascribed "experimental testable reality" in the sense of having a lifetime greater than 10^{-4} s. He specifically disregarded effects due to the shell structure of the atomic nucleus. Referring to recent developments in stellar nucleosynthesis Werner and Wheeler (1958) briefly speculated that nuclei with mass number as high as $A = 500$ might be produced in astrophysical processes by means of neutron capture. "There seems to be no difficulty on this score in considering the production of superheavy nuclei," they wrote.

A decade later experimentalists began looking for SHEs produced in artificial nuclear reactions or existing in nature. Early as well as later researchers sometimes defined SHEs as elements located in or near the theoretically predicted "island of stability" around $(Z, N) \cong (114, 184)$ which will be considered in Sect. 2.2. Thus, Seaborg and Loveland (1990, p. 287) suggested that the term SHE should be associated with "an element whose lifetime is strikingly longer than its neighbors in the chart of the nuclides." The same terminology was used in Thompson and Tsang (1972), one of the earliest SHE review articles. Apart from Werner and Wheeler (1958) the first scientific papers with "superheavy" or "super-heavy" elements in the title appeared in the 1960s. According to the *Web of Science*, today (November 2017) the cumulative number of such papers has grown to 1,996 with 251 of them using the hyphenated spelling "super-heavy" (Fig. 1.1).

Theories and experiments on SHEs have attracted massive scientific interest for about half a century, resulting in thousands of papers and numerous conferences and symposia. Compared to this activity, strikingly little has been written on the subject from a historical perspective and even less from a philosophical perspective. One might imagine that historians of physics and chemistry have taken up the subject, but this is not the case. Remarkably, the comprehensive database established by the scholarly journal *Isis* (https://data.isiscb.org), a standard bibliographic tool among professional historians of science, lists just a single publication on SHEs, namely Kragh (2013). The classic work on the discovery of elements by Weeks and Leicester (1968), covers the transuranic elements but only up to about $Z = 103$. This is also where Seaborg (1963), a semi-popular book on synthetic elements, ends. Aspects of the more modern development are included in Fontani et al. (2015).

This is not to say that historical information about modern SHE research is lacking, only that the subject has not yet entered professional history of science. There exist several books and valuable scientific reviews written by participants in

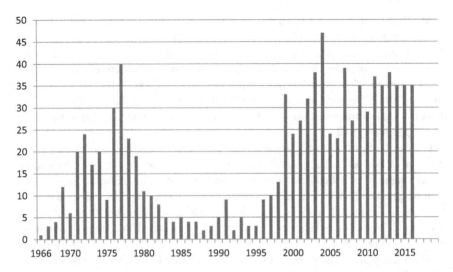

Fig. 1.1 Number of papers including "superheavy" or "super-heavy" in the title. Data based on the web of science

the development and mainly aimed at the SHE community broadly conceived. Monographs include Seaborg and Loveland (1990), Hoffman et al. (2000), and Hofmann (2002). Among the informative review articles are Armbruster and Münzenberg (2012) and Herrmann (2014), both of them focusing on the Darmstadt group in SHE research. The important work done by Russian and other scientists in the Dubna research centre is reviewed in Flerov and Ter-Akopian (1981) and, for the later development, in Oganessian et al. (2004).

There is little doubt that the subject of SHE research, if only properly studied and contextualised, is of interest from the point of view of history, philosophy and sociology of science. As Scerri (2012, p. 336) mentions, the synthesis of super-heavy elements "has raised some new philosophical questions regarding the status of the periodic law." And this is by far the only question of a philosophical and meta-scientific nature that the subject raises. For one thing, SHE research offers a new window to the current relationship between chemistry and nuclear physics. For another thing, it provides case studies illuminating key concepts such as prediction, discovery, and reproducibility of experiments; and, related to sociology of science, fraud and scientific misconduct.

The SHE story also problematizes the crucial notion of what constitutes a chemical element (Kragh 2017). Can one reasonably claim that superheavy nuclei exist in the same sense that the element oxygen exists? Apart from this ontological question, SHE research also involves the epistemic question of how knowledge of a new SHE is obtained and what the criteria for accepting discovery claims are. And, what is equally important, *who* are responsible for the criteria and for evaluating discovery claims? These questions will be considered in Chap. 7.

Table 1.1 Used abbreviations

ACS	American Chemical Society
CIAAW	Commission on Isotopic Abundance and Atomic Weights
CNIC	Commission on Nomenclature of Inorganic Chemistry
GANIL	Grand Accélérateur National d'Ions Lourds
GSI	Gesellschaft für Schwerionenforschung
ICAW	International Committee on Atomic Weights
ICCE	International Committee on Chemical Elements
IUC	International Union of Chemistry
IUPAC	International Union of Pure and Applied Chemistry
IUPAP	International Union of Pure and Applied Physics
JINR	Joint Institute for Nuclear Research
JWP	Joint Working Party
LBNL	Lawrence Berkeley National Laboratory
LLNL	Lawrence Livermore National Laboratory
SHE	Superheavy element
TWG	Transfermium Working Group

The history of transuranic elements is much too rich to be covered in its entirety and consequently I only discuss select parts of it. The essay starts with a brief account of the early period from the 1890s to the 1940s involving theoretical speculations as well as the discovery of the first elements beyond uranium. In Chap. 2 parts of the later history is recounted, including attempts to find SHEs in nature. Chapter 3 deals with the background of SHE research and the development up to about 1970, to be followed in Chap. 4 by a discussion of a couple of controversial element discovery claims and, in Chap. 5, the more extended controversy over the so-called transfermium elements. Aspects of the more recent work on elements with $Z > 110$ are dealt with in Chap. 6. The essay ends with a chapter of a more general character in which I discuss questions of a philosophical and sociological nature relating to modern SHE research. This chapter is in part of a more provisional and speculative nature.

SHE is only one among several abbreviations appearing in this essay. The more important ones are listed in Table 1.1.

1.2 Early Speculations

Mendeleev's periodic system of the elements was based on the atomic weight M, which until the early 1910s was universally considered the defining parameter of a chemical element (Kragh 2000). Although there were a few speculations that atoms of the same element might differ in weight, such as proposed by the British chemist William Crookes in 1886, these were not generally accepted. A drastic and most fruitful change occurred with the redefinition of a chemical element in terms of the atomic number, as first proposed by Antonius van den Broek in 1912 and turned into an operational method by Henry Moseley in 1913. Moseley's work based on X-ray spectroscopy was singularly important as it provided a unique method of determining the atomic number. According to Frederick Soddy, Nobel Prize laureate and pioneer of isotopic chemistry, "For the first time Moseley had called the roll of the elements and we could now say definitely the number of possible elements between the beginning and the end, and the number that still remained to be found" (Childs 1998, p. 39). Soddy seems to have taken for granted, incorrectly as it turned out, that uranium marked the end of the periodic table.

In 1925 it was proposed to honour Moseley with the name of element 43, to be called "moseleyum," but the proposal fell on deaf ears (Swinne 1925; Fontani et al. 2015, pp. 286–288). The introduction of the atomic number and its interpretation as the positive charge of the atomic nucleus ruled out earlier speculations of elements lighter than hydrogen with $Z = 1$. On the other hand, the replacement of M with Z as the ordering parameter did not change the situation with regard to the possible existence of elements heavier than uranium. Obviously there could not be less than one proton in the nucleus, but why not more than 92? Or, according to the proton-electron picture of the atomic nucleus accepted by physicists in the 1920s, why could there be no more than 238 protons?

Even before the introduction of the atomic number there were several speculative attempts to extend the periodic table with hypothetical elements heavier than uranium; or, alternatively, to justify theoretically why $M \cong 240$ was a maximum atomic weight (Quill 1938; Kragh 2013). These attempts were in part inspired by two important *fin de siècle* discoveries, one being the discovery of argon and other inert gases and the other the slightly later discovery of radioactivity.

Consider the Danish chemist Julius Thomsen, a founder of rational thermochemistry, who in April 1895 suggested that transitions from the halogens to the alkali metals went through a group of intermediate elements similar to argon as a bridge between chlorine and potassium. There exist, he said, "inactive elements with atomic weights 4, 20, 36, 84, 132, 212 and 292, the electrochemical character of which is indifference and the valence of which is zero" (Thomsen 1895; Kragh 2016, pp. 303–306.). His predictions as given in Table 1.2 were surprisingly good except perhaps for argon. Thomsen argued that thorium and uranium belonged to a period of 32 elements which ended with an inert gas of $M \cong 292$, very close to the atomic mass assigned to element 118 discovered more than a century later.

Table 1.2 Thomsen's 1895 prediction of a group of inert gases

Symbol	He	Ne	Ar	Kr	Xe	Rn	Og
Atomic number	2	10	18	36	54	86	118
Atomic weight, Thomsen	4	20	36	84	132	212	292
Atomic weight, modern	4.0	20.2	39.9	83.8	131.3	222	(294)
Discovery	1895	1898	1894	1898	1898	1910	2016

While Thomsen shared with all contemporary chemists the belief that the order of the periodic table was determined by the atomic weight, the Swedish physicist Johannes Rydberg adopted the new atomic number in his 1913 explanation of the table. Rydberg derived an element of group zero corresponding to the successor of radon and with atomic number 118, but he did not assign an atomic weight to it (Rydberg 1914, p. 606). In a purely formal way he also derived even heavier group-zero elements of $Z = 168, 218, 290, 362$ and 460. These hypothetical elements were superheavy indeed. Although Rydberg thought that they fitted into his extended version of the periodic table, apparently he did not conceive them as real elements existing in nature.

At the time of Thomsen's prediction radioactivity was still in the future, but not for long. The new phenomenon caused concern among the chemists, for could a spontaneously disintegrating substance be regarded a *bona fide* chemical element? A few chemists thought that they might have found radioactive elements heavier than uranium, but their suggestions were not substantiated. Worth of brief notice is the American chemist Charles Baskerville, at the University of North Carolina, who in investigations of thorium salts in 1904 found a fraction whose atomic weight he determined to $M = 255.8$ (Baskerville 1904; Fontani et al. 2015, pp. 192–195). According to Baskerville, he had discovered "carolinium," a new quadrivalent element heavier than uranium. He wisely avoided placing his transuranic element in the periodic table. However, carolinium (so named after North Carolina) was but a brief parenthesis in the history of chemistry. It failed to win recognition in the chemical community and shared with numerous other spurious elements that it was a name without a reality. Baskerville's element is worth mention only because it was the first empirically based discovery claim of a transuranic element.

The large majority of chemists refrained from speculating about transuranic elements and the limits of the periodic system. After all, chemistry was an empirical science dealing what was known and not with what was not known. In a wide-ranging book on the chemical elements published in 1910, the British chemist William Tilden considered the question. Undoubtedly expressing the majority view of the time, he wrote as follows (Tilden 1910, pp. 56–57):

> There is nothing in theory to preclude the addition of new substances to either extremity of the series. However, ... the belief that radio-activity is occasioned by the instability of the larger atoms ... leads to the suspicion, if not the conviction, that the series is limited by this instability, which, as far as is at present known, is associated with an atomic weight approximating to 240.

An advocate of the then fashionable view of chemical evolutionism, Tilden believed that all the elements had once been formed by condensation processes of primitive atoms. But he left open the possibility that the "primal stuff" might be "uranium itself or something of still higher atomic weight existing in small quantity in uranium."

1.3 Atomic Calculations

With the development of the Bohr-Sommerfeld quantum theory of atomic structure it became possible, for the first time, to address the question of the upper limit of the periodic table by means of strictly scientific reasoning. In Niels Bohr's theory of the periodic system he assigned electron configurations in terms of two quantum numbers (n, k) to all chemical elements from hydrogen to uranium, where n is the principal and k the azimuthal quantum number. According to the old orbital quantum theory the quantum numbers could attain the values

$$n = 1, 2, 3, \ldots \quad \text{and} \quad k = 1, 2, 3, \ldots, n$$

The k quantum number of the old theory translates into the standard notation of modern quantum mechanics by

$$l = k - 1 = 0, 1, 2, \ldots$$

On a few occasions Bohr even went into the *terra incognita* of transuranic elements, extending the configurations to hypothetical elements with $Z > 92$. In 1922, first in a lecture series in Göttingen and then in his Nobel Prize lecture in Stockholm, he predicted the configuration of $Z = 118$, stating that it would be an inert gas with properties similar to radon (Bohr 1977, p. 405; Kragh 2013). Written as the population of electrons in the various shells given by the principal quantum number he argued that element 118 would have the structure

$$2, 8, 18, 32, 32, 18, 8$$

Bohr did not really believe that such a superheavy element existed and merely presented the case of $Z = 118$ as an illustration of the power of his atomic theory. Remarkably, today the electron configuration of the element oganesson is predicted to be just the same as Bohr's proposal (Nash 2005). The Danish physicist was among the first to realise that the end of the periodic system was given by the radioactive properties of the atomic nucleus and not only by the electron system. As Bohr (1922, p. 112) wrote, "nuclei of atoms with a total charge greater than 92 will not be sufficiently stable to exist under conditions where the elements can be observed."

All the same, Bohr and a few other quantum physicists investigated theoretically the upper limit of the periodic system in terms of the electron structure. In publications from 1923 and 1924 Bohr and his Norwegian assistant Svein Rosseland argued that an electron moving in an (n, k) orbit would fall into the nucleus if the ratio Z/k exceeded the inverse fine-structure constant:

$$\frac{Z}{k} \geq \frac{hc}{2\pi e^2} = \frac{1}{\alpha} \cong 137$$

For $k = 1$, this implies that $Z \leq 137$, which "offers a hint toward an understanding of the limitation in the atomic number of existing elements" (Bohr 1923, p. 266). Based on more sophisticated theory and by taking into account the relativistic mass increase of the electrons Arnold Sommerfeld in Munich reached the same conclusion. He proved that if $k < \alpha Z$ the electron would spiral around the nucleus, approaching it with almost the speed of light. For $k = 1$, $Z = 137$ would therefore be the limit allowing elliptic orbits.

The question of the number of chemical elements was also reconsidered by other physicists relying on arguments somewhat similar to Bohr's and Sommerfeld's. Among them were three German physicists, Walther Kossel in Kiel, Walter Gordon in Hamburg, and Erwin Madelung in Frankfurt. Madelung (1936, p. 360) suggested electron configurations for hypothetical elements in the interval $92 < Z < 104$. In all of these calculations the charge of the nucleus was assumed to be concentrated in a point.

Of greater interest still is the little known work of Richard Swinne, a privately employed German physicist and engineer with an interest in the missing elements of the periodic table (Swinne 1925; Fontani et al. 2015). In a review paper of 1926 he gave serious consideration to the possible existence of "transuranic" elements, a name he may have coined. Swinne supported Bohr's suggestion that the apparent absence of elements heavier than uranium was probably due to the instability of nuclei with $Z > 92$. He went beyond Bohr by arguing from considerations of radioactivity that nuclear instability did not simply increase with the atomic number: "Long-lived elements should first appear from about $Z = 98$ to $Z = 102$ and then again from about $Z = 108$ to $Z = 110$, but in between these there should be only short-lived elements" (Swinne 1926, p. 212). He had come to this hypothesis as early as 1914, in a lecture given to Heidelberg Chemical Society.

Although Swinne's reasons for the prediction were not entirely clear, he anticipated what much later became known as the idea of a transuranic "island of stability" (Stradins et al. 1987). He also went beyond Bohr and other physicists by suggesting for the first time detailed electron configurations for elements with $92 < Z < 108$. For $Z = 104$ (the later rutherfordium) he suggested that it had the shell structure

$$2, 8, 18, 32, 18, 18, 8$$

and thus might be the true homologue to radon instead of Bohr's $Z = 118$. As seen in retrospect, Swinne's ideas concerning atomic structure and relatively long-lived transuranic elements were interesting, but they failed to attract attention and were soon forgotten. They had no impact at all on the later development of the nuclear properties of very heavy elements.

Swinne further considered the hypothesis that the absence of transuranic elements in the crust of the Earth might be due to geochemical circumstances that

Table 1.3 Goldschmidt's predictions of ionic radii of transuranic elements

Z	93	94	95	96	97	98
Modern symbol	Np^{4+}	Pu^{4+}	Am^{4+}	Cm^{4+}	Bk^{4+}	Cf^{4+}
Estimated radius (Å)	1.03	1.01	0.99	0.98	0.96	0.94
Modern radius (Å)	1.01	1.00	0.99	0.99	0.97	0.96

allowed them to exist only in the interior of our globe. In this connection he referred to the eminent Norwegian mineralogist Goldschmidt (1924) who had suggested that the elements with $Z = 94$, 95 and 96 constituted a "neptunium group" homologous to the platinum group. According to Goldschmidt, traces of these unknown elements might possibly be found in minerals of platinum and iridium. In the 1930s Goldschmidt speculated on the chemical properties of transuranic elements, arguing that they belonged to what he called the "thorides" and are better known as the actinides (Mason 1992, p. 76; Goldschmidt 1954, pp. 432–434). As shown by Table 1.3, his estimates of the radii of the quadrivalent ions were remarkably precise.

Swinne shared the belief that some of the relatively long-lived transuranic elements might exist on Earth, possibly in its solid core or in iron meteorites. Moreover, he speculated that the elements might be found in the iron-rich so-called polar dust which supposedly was of cosmic origin and had been locked up in the Greenlandic ice cap. According to the "radioactive dust hypothesis" as expounded in Swinne (1919), the penetrating and at the time enigmatic cosmic rays were of solar origin and in part consisting of radioactive nuclei expelled from the Sun. Swinne (1926) had the opportunity to examine samples of ice from Greenland and, he suggested, "The X-ray examination indicated with some uncertainty the presence of $Z = 108$." Unable to verify the result he did not claim that he had actually discovered the element. In a later paper Swinne (1931) returned to the subject, now suggesting that hypothetical elements of $Z \cong 100$ and $Z \cong 108$ might be relatively long-lived and possibly exist in nature. He now accepted Bohr's view that element 118 would be homologous to radon, from which he went on proposing electron structures for the even higher homologues $Z = 168$ and $Z = 218$.

Swinne was not alone in assuming that terrestrial radioactivity might somehow be induced by cosmic rays. The German-American radiochemist Aristid Grosse held somewhat similar ideas of what he called "cosmic radio-elements."[1] Indeed, among German scientists there was in the inter-war period much interest in transuranic elements whether in a terrestrial or cosmic context. To mention but one more example, the Nobel Prize-rewarded physical chemist Walther Nernst believed that "the sources of the energy of the fixed stars must be looked for in radio-active

---

[1]Grosse (1934) to which Goldschmidt (1954), p. 434 referred positively. For his early work on the still hypothetical elements $Z = 93$ and 94, see Grosse (1935). Much later, after several of the transuranic elements had been synthesised, he made detailed calculations of the physical and chemical properties of $Z = 118$, which he called eka-emanation. See Grosse (1965).

elements which are of higher atomic weight than uranium" (Nernst 1921, 1928, p. 137). He also thought that the hypothetical elements might explain the high energies of the cosmic rays. Although Nernst's advocacy of celestial superheavy elements was unorthodox, he was not the only reputed scientist of the period who thought that astronomy and cosmology required elements heavier than uranium.

The famous British astronomer James Jeans argued in publications from the late 1920s that the energy emitted by massive stars was mainly due to the disintegration of transuranic elements with atomic number 95 or even higher (Jeans 1928). However, the large majority of physicists and astronomers dismissed his hypothesis and with the understanding in the late 1930s of stellar energy in terms of nuclear fusion processes superheavy elements largely disappeared from the astrophysical literature (but see Sect. 6.3).

1.4   From Fermi to Seaborg

The first artificial transformation of one chemical element into another element dates from 1919, when Ernest Rutherford concluded from experiments that nitrogen irradiated by alpha particles changed into a carbon isotope of mass number 13. It took a couple of years until it was realised that the product was in fact O-17 and not C-13. The experiments on nuclear transformation made during the 1920s were mostly conducted with projectiles from naturally occurring alpha emitters and for the most part they were restricted to target elements from the lower part of the periodic table.

The possibility of manufacturing transuranic elements in the laboratory only became a possibility after 1932, when the neutron was discovered and the atomic nucleus reconceptualised as a proton-neutron rather than a proton-electron composite. Enrico Fermi and his group in Rome, including Edoardo Amaldi, Emilio Segré and Oscar D'Agostino, studied systematically neutron reactions with all the elements of the periodic system, paying particular attention to the heaviest elements. Famously, when they bombarded uranium with slow neutrons they obtained results which suggested that they had obtained an element of atomic number greater than 92. Indeed, for a time Fermi and his group believed that they had produced elements with $Z = 93$ and $Z = 94$. When Fermi was awarded the 1938 Nobel Prize in physics it was in part because he had "succeeded in producing two new elements, 93 and 94 in rank number. These new elements he called Ausenium and Hesperium."[2] Fermi, who was nominated for both the physics and chemistry prize, only put forward the tentative names in 1938 (Segré 1980, p. 205). The two elements were believed to have been formed as

---

[2]Presentation speech by Henning Pleiel of 10 December 1938. Online as https://www.nobelprize.org/nobel_prizes/physics/laureates/1938/press.html.

$$\ce{^{238}_{92}U} + \ce{^{1}_{0}}n \rightarrow \ce{^{239}_{92}U} \rightarrow \ce{^{239}_{93}Ao} + \beta^- \rightarrow \ce{^{239}_{94}Hs} + 2\beta^-$$

However, it quickly turned out that the Nobel Foundation's support of ausenium and hesperium was an embarrassing mistake. The physics prize of the following year was awarded to Ernest Lawrence for his invention of the cyclotron, and also in this case did the prize motivation refer to the "artificial radioactive elements" resulting from the invention.

Although this essay is about artificially made transuranic elements it is worth recalling that the first new element ever produced in the laboratory was sub-uranic. In 1937 Segré and his collaborator, the Italian mineralogist Carlo Perrier, analysed plates of molybdenum irradiated with deuterons and neutrons from the Berkeley cyclotron. They were able to identify two isotopes of element 43 (Tc-95 and Tc-97) for which they proposed the name technetium ten years later. There had earlier been several unconfirmed claims of having detected the element—which for a time was known under the names "masurium" or "illinium"—in natural sources (Scerri 2013, pp. 116–143). With their 1937 synthesis Segré and Perrier soon became recognised as the true discoverers. Segré is also recognised as the co-discoverer, together with Dale Corson and Kenneth MacKenzie, of element 85 which was produced in Berkeley in 1940 by bombarding Bi-209 with alpha particles. In 1947 they suggested the name astatine for it. Tiny amounts of astatine exist in nature, and also in this case there were previous claims of having identified the very rare element (Thornton and Burdette 2010).

The early attempts to produce transuranic elements are thoroughly described in the literature, not least because of their intimate connection to the discovery of the fission of the uranium nucleus (Heilbron and Seidel 1989, pp. 456–464; Sime 2000; Seaborg 2002). For a detailed contemporaneous review of the confused situation shortly before the discovery of fission, see Quill (1938). We only need to point out that it was investigations of fission fragments that led Edwin McMillan and Philip Abelson, at the Berkeley Radiation Laboratory, to conclude that they had detected element 93 as a decay product of the neutron-produced isotope U-239. This first transuranic element was announced in *Physical Review* on 27 May 1940, at a time when the self-imposed publication ban caused by the war was still not effective. According to *Oakland Post Inquirer*, a local newspaper, the discovery might "conceivably prove more influential in the destiny of the world than any single battle of the current World War" (Heilbron and Seidel 1989, p. 459). Keeping to the astronomical analogy Uranus-uranium, the new element was called neptunium, chemical symbol Np.

Even more importantly, in late 1940 Glenn Seaborg and collaborators prepared the isotope Np-238 by bombarding uranium with deuterons:

$$\ce{^{238}_{92}U} + \ce{^{2}_{1}H} \rightarrow \ce{^{238}_{93}Np} + 2\ce{^{1}_{0}}n$$

They identified the daughter product as an element of $Z = 94$:

$$^{238}_{93}\text{Np} \rightarrow ^{238}_{94}\text{Pu} + \beta^-$$

The Berkeley group also produced the more important isotope of mass number 239, soon showed to be fissionable. The discovery paper was submitted on 7 March 1941, but as a result of the war it only appeared in print more than five years later (Seaborg et al. 1946). In the words of Seaborg, the final discovery event was as follows: "On the stormy night of February 23, 1941, in an experiment that ran well into the next morning, [Arthur] Wahl performed the oxidation that gave them proof that what they had made was chemically different from all other known elements" (Hoffman et al. 2000, p. 34).

The name of the new element, plutonium (Pu), first appeared in a classified memorandum of March 1942 and only six years later in a scientific paper. Following the names of uranium and neptunium, it was named after the next planet, Pluto, which at the time was the outermost planet in the solar system. (This position was lost in 2006 when Pluto was degraded to a dwarf planet). Needless to say, it quickly turned out that plutonium was more than just another exotic element of interest only to the scientists. The inhabitants of Nagasaki experienced that on 9 August 1945. Several of the transuranic elements have been made in visible quantities and a few of them, including long-lived isotopes of curium and americium, have even found applications in science and industry. Plutonium is unique by being the only synthetic element produced in very large quantities. It is estimated that today the world stockpile of the element is about 400 tons corresponding to about 10^{30} atoms (Armbruster and Münzenberg 1989). The long half-life of plutonium, namely 2.4×10^4 years for Pu-239 and 8×10^7 years for Pu-244, means that the element is not just an ephemeral visitor on Earth but will remain with us for thousands of years to come. Pu-238, a powerful alpha emitter with a half-life of 88 years, is used for thermoelectric batteries in space missions.

The early history of transuranic elements was completely dominated by a group of Californian chemists and physicists led by Seaborg and Albert Ghiorso. While Seaborg was trained in chemistry and had worked under the eminent chemist Gilbert N. Lewis, Ghiorso's background was in electrical engineering and physics instrumentation. When Seaborg moved to Chicago in the early 1940s to work in the Manhattan Project, he invited the three years younger Ghiorso to join him. Elements 95 and 96 were first identified in 1944 at the Metallurgical Laboratory in Chicago and named americium (Am) and curium (Cm), respectively.[3] Whereas the first element was identified as a beta decay product of Pu-241, the latter was produced by bombarding Pu-239 with alpha particles in the Berkeley cyclotron:

---

[3]See Seaborg (1994b) for a lively account of the two elements' naming history, and Kostecka (2008) for the discovery and applications of americium. The name "curium" in honour of Marie and Pierre Curie was the first time an element had been named after a scientist.

$$^{239}_{94}\text{Pu} + {}^{4}_{2}\text{He} \rightarrow {}^{242}_{96}\text{Cm} + {}^{1}_{0}n$$

The product was identified by measuring the characteristic energy of the alpha particle emitted during the decay of Cm-242 to Pu-238. It turned out that the two elements had nearly identical chemical properties and that it was therefore very difficult to separate them. The difficulty led one member of the discovery team to refer to them in jest as "pandemonium" and "delirium" (Seaborg and Loveland 1990, p. 21).

It was the problems with synthesising curium and separating it chemically from americium that led Seaborg to propose, first in secret report of 1944, a modified form of the periodic table sometimes known as the Mendeleev-Seaborg table. The modification consisted in adding what Seaborg called an actinide series with elements from $Z = 89$ to 103. The first version of the periodic table including Seaborg's actinide series appeared in *Chemical and Engineering News* at the end of 1945 (Seaborg 1994a, pp. 145–148; Seaborg 1995). Not only did Seaborg introduce the actinide series, in later publications he also suggested the transactinide series (Z from 104 to 121) and even the superactinide series (Z from 122 to 153).

After the war followed the discovery of $Z = 97$ (berkelium, Bk) and $Z = 98$ (californium, Cf) which were announced in 1950 and also obtained by means of a beam of cyclotron-produced alpha particles. The reactions were

$$^{241}_{95}\text{Am} + {}^{4}_{2}\text{He} \rightarrow {}^{243}_{97}\text{Bk} + 2{}^{1}_{0}n \quad \text{and} \quad {}^{242}_{96}\text{Cm} + {}^{4}_{2}\text{He} \rightarrow {}^{245}_{98}\text{Cf} + {}^{1}_{0}n$$

The identification of the californium isotope was accomplished with a total of some 5,000 atoms with a half-life of 44 min. Although this was a very small amount—"substantially smaller than the number of students attending the University of California" (Seaborg 1963, p. 19)—it was much greater than in future element discoveries. In 1951, at a time when six transuranic elements had been added to the periodic system, Seaborg and McMillan were awarded the Nobel Prize in chemistry "for their discoveries in the chemistry of the transuranium elements." The new elements were unstable, but still they were more stable than Fermi's ausenium and hesperium.

Incidentally, the Seaborg-McMillan prize of 1951 is the only Nobel Prize awarded for research in transuranic or superheavy elements. Notice that the prize was in chemistry and not in physics, indicating that discoveries of new elements traditionally belong to the first science (see also Sect. 7.1). While Seaborg was a chemist, MacMillan was a physicist. As MacMillan (1951) pointed out in his Nobel lecture, "in spite of what the Nobel Prize Committee may think, I am not a chemist." He was far from the only physicist to be awarded a Nobel Prize in chemistry. To mention but a few others, also Rutherford, Marie Curie, Francis Aston, Peter Debye, and Gerhard Herzberg were chemistry Nobel laureates. See the table in Kragh (1999, p. 432), to which may be added Gerhard Ertl and Stefan Hell, laureates of 2007 and 2014, respectively.

References

Anon: Superheavy element 94 discovered in new research. Science News Lett. **37** (22), 387 (1940)

Armbruster, P., Münzenberg, G.: Creating superheavy elements. Sci. Am. **144**, 66–72 (1989)

Armbruster, P., Münzenberg, G.: An experimental paradigm opening up the world of superheavy elements. Eur. Phys. J. H **37**, 327–310 (2012)

Baskerville, C.: Thorium, carolinium, berzelium. J. Am. Chem. Soc. **26**, 922–941 (1904)

Bohr, N.: The Theory of Spectra and Atomic Constitution. Cambridge University Press, Cambridge (1922)

Bohr, N.: Linienspektren und Atombau. Ann. Phys. **71**, 228–288 (1923)

Bohr, N.: Niels Bohr. In: Nielsen, J.R. (ed.) Collected Works, vol. 4. North-Holland, Amsterdam (1977)

Childs, P. E.: From hydrogen to meitnerium: naming the chemical elements. In: Thurlow, K. (ed.) Chemical Nomenclature, pp. 27–66. Kluwer Academic, Dordrecht (1998)

Flerov, G.N., Ter-Akopian, G.M.: The physical and chemical aspects of the search for superheavy elements. Pure Appl. Chem. **53**, 909–923 (1981)

Fontani, M., Costa, M., Orna, M.V.: The Lost Elements: The Periodic Table's Shadow Side. Oxford University Press, Oxford (2015)

Gamow, G.: Concerning the origin of chemical elements. J. Wash. Acad. Sci. **32**, 353–355 (1942)

Goldschmidt, V. M.: Geochemische Verteilungsgesetze der Elemente 2. Beziehungen zwischen den geochemischen Verteilungsgesetzen und dem Bau der Atome. Norske Videnskabs-Akademien, Skrifter, Mat. Nat. Klasse (4) (1924)

Goldschmidt, V.M.: Geochemistry. Clarendon Press, Oxford (1954)

Grosse, A.: An unknown radioactivity. J. Am. Chem. Soc. **56**, 1922–1924 (1934)

Grosse, A.: The chemical properties of elements 93 and 94. J. Am. Chem. Soc. **57**, 440–441 (1935)

Grosse, A.: Some physical and chemical properties of element 118 (Eka-Em) and element 86 (Em). J. Inorg. Nucl. Chem. **27**, 509–519 (1965)

Heilbron, J.L., Seidel, R.W.: Lawrence and His Laboratory: A History of the Lawrence Berkeley Laboratory. University of California Press, Berkeley (1989)

Herrmann, G.: Historical reminiscences: The pioneering years of superheavy element research. In: Schädel, M., Shaughnessy, D. (eds.) The Chemistry of Superheavy Elements, pp. 485–510. Springer, Berlin (2014)

Hoffman, D.C., Ghiorso, A., Seaborg, G.T.: Transuranium People: The Inside Story. Imperial College Press, London (2000)

Hofmann, S.: On Beyond Uranium: Journey to the End of the Periodic Table. Taylor & Francis, London (2002)

Jeans, J.: Astronomy and Cosmogony. Cambridge University Press, Cambridge (1928)

Kostecka, K.: Americium—from discovery to the smoke detector and beyond. Bull. Hist. Chem. **33**, 89–93 (2008)

Kragh, H.: Quantum generations: a history of physics in the twentieth century. Princeton University Press, Princeton (1999)

Kragh, H.: Conceptual changes in chemistry: the notion of a chemical element. Stud. Hist. Philos. Mod. Phys. **31**, 435–450 (2000)

Kragh, H.: Superheavy elements and the upper limit of the periodic table: early speculations. Eur. Phys. J. H **38**, 411–431 (2013)

Kragh, H.: Julius Thomsen: A Life in Chemistry and Beyond. Royal Danish Academy of Sciences and Letters, Copenhagen (2016)

Kragh, H.: On the ontology of superheavy elements. Substantia **2**, 7–17 (2017)

MacMillan, E. M.: The transuranium elements: Early history. http://www.nobelprize.org/nobel_prizes/chemistry/laureates/1951/mcmillan-lecture.pdf (1951)

Madelung, E.: Die Mathematischen Hilfsmittel des Physikers. Springer, Berlin (1936)

Mason, B.: Victor Moritz Goldschmidt: Father of Modern Geochemistry. The Geochemical Society, San Antonio, TX (1992)

Nash, C.: Atomic and molecular properties of elements 112, 114, and 118. J. Phys. Chem. A **109**, 3493–3500 (2005)

Nernst, W.: Das Weltgebäude im Lichte der neueren Forschung. Springer, Berlin (1921)

Nernst, W.: Physico-chemical considerations in astrophysics. J. Frankl. Inst. **206**, 135–142 (1928)

Oganessian, YuT, et al.: Heavy element research at Dubna. Nucl. Phys. A **734**, 109–123 (2004)

Quill, L.L.: The transuranium elements. Chem. Rev. **23**, 87–155 (1938)

Rydberg, J.R.: Recherches sur le système des éléments. Journal de Chimie et Physique **12**, 585–639 (1914)

Scerri, E.: The Periodic Table: Its Story and Its Significance. Oxford University Press, Oxford (2007)

Scerri, E.: The periodic table. In: Woody, A., Hendry, R., Needham, P. (eds.) Philosophy of Chemistry, pp. 329–338. North-Holland, Amsterdam (2012)

Scerri, E.: A Tale of 7 Elements. Oxford University Press, Oxford (2013)

Seaborg, G.T.: Man-Made Transuranium Elements. Prentice-Hall, Englewood Cliffs (1963)

Seaborg, G.T.: Modern Alchemy: Selected Papers of Glenn T. Seaborg. World Scientific, Singapore (1994a)

Seaborg, G.T.: Terminology of the transuranium elements. Terminology **1**, 229–252 (1994)

Seaborg, G.T.: Transuranium elements: past, present, and future. Acc. Chem. Res. **28**, 257–264 (1995)

Seaborg, G.T.: Prematurity, nuclear fission, and the transuranium actinide elements. In: Hook, E.B. (ed.) Prematurity in Scientific Discovery, pp. 37–45. University of California Press, Berkeley (2002)

Seaborg, G.T., Loveland, W.D.: The Elements Beyond Uranium. Wiley, New York (1990)

Seaborg, G.T., Wahl, A.C., Kennedy, J.W.: Radioactive element 94 from deuterons. Phys. Rev. **69**, 367 (1946)

Segré, E.: From X-Rays to Quarks: Modern Physicists and Their Discoveries. W. H. Freeman, San Francisco (1980)

Sime, R.L.: The search for transuranium elements and the discovery of nuclear fission. Phys. Perspect. **2**, 48–62 (2000)

Stradins, J.P., Trifonow, D.N., Pijola, S.: Die Evolution der Idee von 'Inseln Relativer Stabilität' der Chemischen Elemente. D.A.V.I.D. Verlagsgesellschaft, Berlin (1987)

Swinne, R.: Zum Ursprung der durchdringenden Höhenstrahlung. Naturwissenschaften **7**, 529–530 (1919)

Swinne, R.: Zwei neue Elemente: Masurium und Rhenium. Zeitschrift für Technische Physik **6**, 464–465 (1925)

Swinne, R.: Das periodische System der chemischen Elemente im Lichte des Atombaus. Zeitschrift für Technische Physik **7**(166–180), 205–216 (1926)

Swinne, R.: Zur Periodizität der Atomkerne. Wissenschaftliche Veröffentlichungen aus dem Siemens-Konzern **10**, 137–147 (1931)

Thompson, S.G., Tsang, C.F.: Superheavy elements. Science **178**, 1047–1055 (1972)

Thomsen, J.: Über die mutmasslische Gruppe inaktiver Elemente. Z. Anorg. Chemie **8**, 283–288 (1895)

Thornton, B.F., Burdette, S.C.: Finding eka-iodine: discovery priority in modern times. Bull. Hist. Chem. **35**, 81–85 (2010)

Tilden, W.A.: The Elements: Speculations as to their Nature and Origin. Harper & Brothers, London (1910)

Weeks, M.E., Leicester, H.M.: Discovery of the elements. J. Chem. Educ. Easton, PA (1968)

Werner, F.G., Wheeler, J.A.: Superheavy nuclei. Phys. Rev. **109**, 126–143 (1958)

Wheeler, J. A.: Nuclear fission and nuclear stability. In: Pauli, W. (ed.) Niels Bohr and the Development of Physics, pp. 163–184. Pergamon Press, London (1955)

Chapter 2
Transuranic Alchemy

Abstract The history of superheavy elements (SHEs) cannot be cleanly separated from the earlier history of transuranic elements with atomic numbers smaller than 104. By the late 1960s elements up to this place in the periodic table, including lawrencium of $Z = 103$, had been discovered although in a few cases the discovery claims were controversial. In this period the group of Berkeley nuclear scientists faced the first serious competition from scientists in Dubna in the Soviet Union. The beginning of proper SHE research was in part inspired by theoretical predictions of a so-called island of stability which also stimulated much work on the possible existence of SHEs in nature. This line of work, culminating in the 1970s, did not result in positive evidence and yet the search for naturally occurring SHEs has continued to this very day. The chapter ends with brief remarks on scientists' motivation for synthesising and examining the very heavy elements.

Keywords Transuranic elements · Superheavy elements · Island of stability
Magic numbers · Albert Ghiorso · Georgii Flerov

Investigations of superheavy elements began in the 1960s, to a large extent motivated by theories of nuclear structure predicting relatively stable nuclei with atomic numbers 110 or more. Much of the early interest in the subject was concerned with finding these very heavy elements in nature, an extensive research activity which involved many physicists, chemists and geologists but led to no positive results. The chapter starts with a brief account of the discovery histories of elements with atomic numbers from 99 to 103.

2.1 Four More Elements

Elements 99 and 100, named einsteinium (Es) and fermium (Fm), were first identified in late 1952, not in a planned experiment but unexpectedly in the debris from the "Mike" test of the American hydrogen bomb in the Pacific on 1 November

H. Kragh, *From Transuranic to Superheavy Elements*, SpringerBriefs in History
of Science and Technology, https://doi.org/10.1007/978-3-319-75813-8_2

1952. Early analysis of the debris indicated that the high neutron flux density of the thermonuclear explosion had caused U-238 to absorb multiple neutrons and form a new plutonium isotope, Pu-244. When Albert Ghiorso and his collaborators in Berkeley examined the radioactive material from the explosion, they were able to detect an isotope emitting alpha particles with the characteristic energy 6.6 meV. They identified the isotope as belonging to element 99. The reaction, which took place by successive neutron capture followed by beta decay, can be summarized as

$$\,^{238}_{92}U + 15\,^{1}_{0}n \rightarrow \,^{253}_{98}Cf + 6\beta^- \text{ and } \,^{253}_{98}Cf \rightarrow \,^{253}_{99}Es + \beta^-$$

A similar reaction was responsible for the synthesis of the 255 isotope of element 100 identified by a 7.1 meV alpha emitter. In 1954 the same isotopes were produced by more conventional means, namely in high-flux nuclear reactors. The discovery team led by Ghiorso only published its findings in 1955, a delay caused by orders from the U.S. military. The team suggested the symbol E for einsteinium, but it was later changed to Es. The priority of the American team was generally recognised and that although the synthesis of an isotope of element 100 was reported in 1954 by a group of Swedish physicists (Atterling et al. 1954). Using the cyclotron at the Nobel Institute for Physics in Stockholm, the Swedes bombarded uranium metal with oxygen ions and found an alpha activity which they ascribed to an isotope of element 100, probably with mass number 250. The Stockholm group did not suggest a name for the element and the following year it became clear that priority belonged to Berkeley.

Also in 1955, the discovery of element 101 (mendelevium, Md) was announced by the Berkeley group, this time by using its cyclotron to irradiate a tiny sample of Es-253 containing only about one billion atoms corresponding to a mass of ca. 10^{-12} g. The projectiles were alpha particles:

$$\,^{4}_{2}He + \,^{253}_{99}Es \rightarrow \,^{256}_{101}Md + \,^{1}_{0}n$$

The delicate experiment led to the production of no more than 17 atoms of Md-256 which were identified one by one by means of the spontaneous fission of Fm-256 atoms produced by electron capture:

$$\,^{256}_{101}Md + e^- (EC) \rightarrow \,^{256}_{100}Fm$$

The half-life of the electron capture decay was estimated to about half an hour, which in later studies was revised to 1.5 h. The number of mendelevium atoms was small indeed, but it was enough to discover the new element. It was the first new element to be identified by means of spontaneous fission, which subsequently became a common detection method for many superheavy elements. The discovery claim of the Ghiorso-Seaborg group was independently confirmed by another Berkeley group in 1958 using the same method but an einsteinium target with a much larger number of Es-253 atoms. Although the latter group did not claim to have contributed to the discovery of element 101, in 1992 it was this group which was officially credited with having discovered the element "with certainty" (Wilkinson et al. 1993,p. 1765).

In a personal account of the discovery, Ghiorso recalled about the name of the new element (Maglich 1972, pp. 245–254): "At this time the political relationship of the U.S. and the U.S.S.R. was not very friendly and it seemed to most people that suggesting such a name would not be a very popular thing to do. Nonetheless we decided that we should rise above politics and honor Mendeleev." The recollection of Seaborg (1994) was the same: "To name an element after a Russian scientist made us vulnerable to criticism during the era of the 'cold war'; this was an unpopular act from the standpoint of some critics in the United States." The name and the chemical symbol Md—changed from the originally proposed Mv—was adopted by IUPAC at its 1957 General Assembly in Paris. Contrary to many of the even heavier elements, in the case of element 101 there were no questions concerning priority. In this respect the discovery history of the element differed significantly from the one of one unit larger, $Z = 102$ or nobelium, which involved several priority controversies. The case of nobelium will be examined in Sect. 3.2.

Whereas the elements from $Z = 93$ to 101 were synthesised in reactions induced by neutrons, deuterons or alpha particles, the next generation of heavy elements involved the use of semi-heavy ions such as B-10, C-13 and O-18. Even heavier elements were eventually produced by means of heavy-ion projectiles as massive as U-238. The elements of atomic number 102 and 103 were the first ones to be produced by heavy-ion reactions.

Following mendelevium it took several more years until the last member of the actinide group, lawrencium of atomic number 103 was discovered, and this time in a more protracted and involved process (Flerov et al. 1991). The first attempts to create the element were made at the Lawrence Berkeley National Laboratory (LBNL) in the late 1950s, but the results were inconclusive as they did not clearly demonstrate that element 103 had been found. Then, on the basis of experiments in early 1961 a small group of Berkeley scientists led by Ghiorso (but not including Seaborg) claimed to have discovered the element.[1] Using a heavy-ion linear accelerator (HILAC) the Ghiorso group bombarded a small target of californium with ions of boron isotopes. By analysing the resulting alpha activity, a new alpha emitter of energy 8.6 meV and half-life ca. 8 s was found and assigned to mass number 257 of element 103. Subsequent experiments proved the assignment to be wrong, as the isotope was most likely Lr-258 produced by

$$^{252}_{98}\text{Cf} + ^{11}_{5}\text{B} \rightarrow ^{258}_{103}\text{Lr} + 5^{1}_{0}n$$

---

[1]Through much of his career, Seaborg was heavily involved in political and administrative work. From 1961 to 1971 he served as chairman of the Atomic Energy Commission (AEC), possibly the most important position in American research policy. During this period he spent most of his time in Washington D.C. When he returned to Berkeley he was appointed professor of chemistry at the University of California.

In this case the discovery claim rested solely on nuclear physics and did not involve chemical analysis. It was followed by a name of the new element, appearing in the title of the paper. "In honor of the late Ernest O. Lawrence, we respectfully suggest that the new element be named lawrencium with the symbol Lw" (Ghiorso et al. 1961). The name was accepted by IUPAC in 1970 but with the element's chemical symbol changed to Lr (Jensen 1971, p. 98).

What appeared to be solid evidence for $Z = 103$ was soon questioned by scientists from the Soviet Union. At the early 1960s Berkeley no longer had monopoly in the manufacture of heavy transuranic elements. Under the leadership of the eminent nuclear physicist Georgii Flerov, the Joint Institute for Nuclear Research (JINR) had been founded in Dubna outside Moscow in 1956 (Blokhintsev 1966). Although JINR was devoted to research in nuclear and high energy physics generally and not specifically to SHE research, it was particularly strong in the latter area. In 1961 the Dubna group was joined by 28-year-old Yuri Oganessian, an Armenian-born physicist and engineer who would become a world leader in SHE research.

The Dubna scientists readily accepted the Berkeley priority for element 101, but not for the element of atomic number two units larger. In fact, they denied the American discovery claim, arguing that it was based either on incorrect experimental data or on errors in interpretation. Many years later, members of the Dubna group concluded that the 1961 discovery paper of Ghiorso and his co-authors "does not contain any serious arguments in favour of the identification of Z." On the contrary, "there is convincing evidence for the priority of the Dubna group in claiming discovery of element 103" (Flerov et al. 1991, pp. 479–480; see also Flerov 1970). The rival claim of the Dubna scientists rested on experiments of 1965 in which they bombarded americium with oxygen ions and reportedly produced the isotope of mass number 256 according to

$$\mathrm{^{243}_{95}Am + ^{18}_{8}O \rightarrow ^{256}_{103}Lr + 5^{1}_{0}}n$$

"It is possible," the authors wrote, "to state unambiguously that $_{103}Lw^{256}$ with a half-life ~ 45 s is synthesised in the reaction" (Donets, Shchegolev and Ermakov 1965). Further experiments in Dubna resulted in the synthesis of also the 255 isotope of what the Russians preferred to call "rutherfordium," their alternative to lawrencium. Without going into further details, generally the discovery story of the element told by the Dubna scientists was quite different from the story told by the Berkeley scientists. The story according to the first group is illustrated in Fig. 2.1.

Whatever doubts there might have been regarding the discovery of element 103 were dispelled when scientists at the Oak Ridge National Laboratory in 1977 detected X-ray lines that unmistakably revealed the 258 isotope. But still the question remained, who was to be credited for the discovery? Only in 1992 did a joint chemistry-physics working group known as the TWG (Transfermium

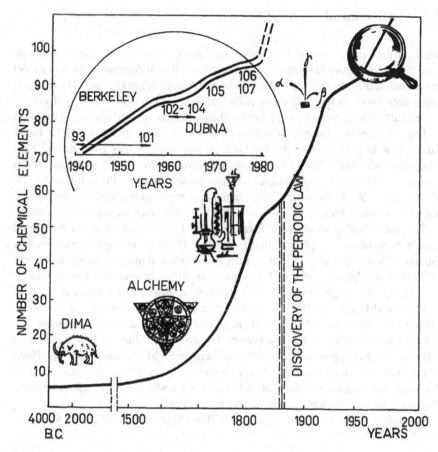

Fig. 2.1 Discoveries of chemical elements as depicted in Flerov and Ter-Akopian (1981). With permission of Pure and Applied Chemistry

Working Group) accept the element, assigning the discovery jointly to Berkeley and Dubna. More about the TWG follows in Sects. 3.1 and 5.2. According to the working group, "In the complicated situation presented by element 103, with several papers of varying degrees of completeness and conviction, none conclusive, and referring to several isotopes, it is impossible to say other than that full confidence was built up over a decade with credit attaching to work in both Berkeley and Dubna" (Barber et al. 1992, p. 484). The name lawrencium was retained and it was reaffirmed that its chemical symbol was Lr and not Lw. At the same occasion it was confirmed that the symbol of einsteinium should be Es and not E, and that the symbol of mendelevium should be Md rather than Mv.

2.2   An Island of Stability?

Much of the impetus for SHE research derives from theories of nuclear structure and in particular from predictions based on the shell or independent-particle model developed in the late 1940s by Maria Goeppert Mayer in the U.S. and Hans Jensen and collaborators in West Germany (Mladenović 1998). In 1963 the two physicists shared half of the physics Nobel Prize for their work, the second half being awarded to Eugene Wigner. However, the idea of a nuclear shell structure can be found earlier, first in a 1933 paper by the German physicists Walter Elsasser and Kurt Guggenheimer. According to Mayer, nuclei with 2, 8, 20, 50, 82, and 126 protons or neutrons were particularly stable, whereas Jensen preferred the numbers 14, 28, 50, 82, and 126. These were "magic numbers" representing closed shells in the nucleus, an idea which mineralogists had anticipated much earlier (Kragh 2000).

The theoretical possibility of a relatively stable element of $Z = 126$ seemed remote from laboratory physics, but in the late 1960s more sophisticated nuclear models indicated that $Z = 114$ rather than $Z = 126$ was a magic number. In a paper of 1966 three physicists at the Dubna centre in Russia argued that the nuclide (Z, N) = (114, 184) might possibly exist (Gareev, Kalinkin and Sobiczewski 1966; Kalinkin and Gareev 2001). The region around this "doubly magic" nucleus was expected to represent nuclei with a relatively long half-life and therefore accessible to experimental study. The region became known as an "island of stability," a term that may first have appeared in the physics literature in Myers and Swiatecki (1966) and which was quickly adopted and promoted by Seaborg. As mentioned in Sect. 1.3, the idea of an island of stability was foreshadowed decades earlier by Richard Swinne (Stradins, Trifonow and Pijola, 1987).

According to the Berkeley physicists William Myers and Wladyslaw Swiatecki (1966, p. 50),

> In order to proceed in a realistic manner with a discussion of the existence and location of possible islands of stability beyond the periodic table the first requirement is the availability of estimates for the location and strength of magic number effects in that region. When such estimates have become available … it will be possible to apply our [our theory] to the prediction of the fission barriers of hypothetical superheavy nuclei.

Another important theoretical paper of the period was due to the Swedish physicist Sven Gösta Nilsson and collaborators who in 1969 calculated the lifetime of nuclides around the region (Z, N) = (114, 184) and discussed the possible production of SHEs in heavy-ion experiments. The calculations of Nilsson et al. (1969) led to the conclusion that the 294 isotope of element 110 had a half-life of approximately 10^8 years and that traces of it might thus have survived in terrestrial nature.

Since the notion of an island of stability was originally introduced, it has changed considerably and is today considered in a different light, not as an island of relatively stable elements placed in a sea of unstable elements but as a region of deformed superheavy nuclei stabilized by shell effects. Apart from (Z, N) = (114,

184) also a magic protonic number of $Z = 164$ has been discussed. However, these more recent developments will not be part of the present study.

It was generally agreed that to reach the island of stability or just come near it, the method to use would be relatively heavy ions bombarding an even heavier target. Work with accelerated heavy ions for nuclear transformations was pioneered at Los Alamos, Yale and Berkeley in the late 1940s, but it took two more decades until heavy-ion accelerators became common in experimental nuclear studies in the U.S. as well as elsewhere. Since then such accelerators have been essential for the synthesis of new SHEs, although SHE research is but one branch among many in the multi-facetted field of heavy-ion science. A detailed historical overview is provided in Bromley (1984).

All known trans-plutonium elements are produced in the laboratory, and yet there is the faint possibility that some of them may exist naturally, either on the Earth, in cosmic rays, or perhaps in the interior of stars or even more exotic places. As mentioned in Sect. 1.2, such speculations were entertained as early as the 1920s. With the idea of an island of stability of mass number $A \sim 300$ including nuclides with half-lives as long as 10^8 years or even longer, interest in the question of naturally occurring SHEs took a new and less speculative turn. Some calculations from the early 1970s indicated that the nuclide $(Z, N) = (110, 184)$ might have a half-life of more than one billion years, as first suggested by Nilsson and co-workers. If so, it would be accessible to experimental study. The hope of the experimenters was to reach the fabled island, if it existed, either by manufacturing the nuclides in the laboratory or by finding them in nature. Both avenues were eagerly followed, often with great vigour and enthusiasm.

2.3 Superheavy Elements in Nature?

As a leading SHE physicist recalled, the predictions from nuclear theory "immediately stirred up a gold-rush period of hunting for superheavy elements in natural samples" (Herrmann 2014, p. 487). Indeed, from about 1970 many researchers began looking for evidence of SHEs in cosmic rays, meteorites, terrestrial ores, or even in samples of lunar matter. In fact, the search for natural transuranic elements had begun at an earlier date. There was in the 1950s, even before the term "superheavy elements" had entered the scientific vocabulary, much interest in naturally occurring elements heavier than plutonium and there were suggestions that some of these elements such as curium might be formed in, for example, explosions of supernovae (Tsaletka and Lapitskii 1960).

In an international Nobel Symposium on SHEs held in Ronneby, Sweden, in June 1974 several of the invited scientists gave addresses on the possibility of finding SHEs in nature. The symposium was attended by 54 scientists of which 22 were from Sweden, 12 from the United States, 9 from Denmark, and 6 from West Germany. In his introduction to the proceedings volume, Nilsson summarized that "most experimentalist authors now appear to conclude that SHE elements are probably *not* present on earth in measurable quantities" (Nilsson and Nilsson 1974). Four years later another SHE symposium took place in Lubbock, Texas, with participation of more than 100 scientists from about 15 different countries, and also at this occasion natural SHEs were given much attention (Lodhi 1978). In an opening address the Oak Ridge physicist Lewin Keller (1978, p. 20) stressed the uncertainty of the SHE enterprise while at the same time appealing to the possibility of serendipitous discoveries:

> We are still groping in a largely unexplored and poorly understood intellectual frontier. Right now this frontier is called superheavy elements; but, of course, it may turn into something else just as exciting. These transformations have occurred in the past as when Fermi's research into transuranium elements turned into the discovery of fission through the inspired and meticulous chemistry of Otto Hahn.

Most of the participants in the Lubbock symposium were Americans, but addresses were also given by scientists from West Germany, Israel, France, Canada, England and Switzerland. No scientists from the USSR and other socialist countries attended this symposium and also not the previous one in Ronneby.

By the early 1980s more than a hundred papers had been published on natural SHEs. Although none was found, in 1971 the nuclear chemist Darleane Hoffman and co-workers were able to detect traces of primordial Pu-244 in old rocks by means of mass spectrometry. Her group detected 2×10^7 atoms of the isotope left over from the formation of our solar system some 5 billion years ago. Plutonium was thus turned into a natural element, at least of a sort (Hoffman et al. 1971; Hoffman, Ghiorso and Seaborg 2000, pp. 38–41).

The favoured method for SHE hunters was to look for tracks due to spontaneous fission, a process which is exceedingly rare in nature but a common decay mode for nuclides with $Z \geq 100$. The decay rate of nuclei suffering spontaneous fission depends on the number of nucleons and roughly varies with A and Z as

$$\log(t_{1/2}) \cong k \frac{Z^2}{A}$$

The first searches for natural SHEs were reported in the late 1960s by research groups from Berkeley, Dubna and elsewhere (Flerov 1970; Thompson and Tsang 1972; Herrmann 1979). Dakowski (1969) cautiously suggested to have found indications of SHEs in meteorites and two years later scientists at the Tata Institute of Fundamental Physics in Mumbai, India, reported "conclusive proof for the presence at some time of super-heavy elements in meteorites" (Bhandari et al. 1971). Although they were unable to determine the atomic number of the suspected element, they conjectured that it might be 114. The alleged conclusive proof quickly turned out to be inconclusive.

In 1973 a team of American scientists reported an extensive investigation of SHEs with Z from 110 to 119 in terrestrial and meteoric minerals, including among their samples 60-million-old sharks' teeth (Stoughton et al. 1973). However, the result of their search was negative. Three years later another American team, led by

Robert Gentry from the Oak Ridge National Laboratory, concluded to have found evidence "of a high level of confidence" for the existence of $Z = 126$ in monazite crystals with an estimated concentration of 10^{-10} g/g. In addition they found weaker evidence for $Z = 116$ and 124 (Gentry et al. 1976). The method that generated such confidence was measurements of X-ray spectra obtained by irradiating samples of radioactive halos with a highly focused beam of protons. The X-ray frequencies seemed to fit nicely with the high atomic numbers.

The announcement in June 1976 of a possible discovery of naturally occurring SHEs caused excitement in the physics community. As *Nature* reported, enthusiastically but prematurely, "the super-heavy elements 116, 124 and 126 have now been detected … and found to be stable" (Hodgson 1976). The findings also found their way to *The New York Times*, where they were described by Walter Sullivan, the newspaper's highly respected science editor. "Unlike earlier observations," Sullivan (1976) concluded, "these involve precise measurements that are open to amplification as well as repetition and verification by other researchers." Indeed, the measurements of Gentry and his co-workers could be and were examined by other physicists, but with the result that the discovery claim soon lost its credibility. More exhaustive and more sensitive experiments unfortunately failed to confirm the presence of element 126 or other SHEs. The characteristic X-ray lines could be explained otherwise and thus the sensational discovery claim vanished after less than a year (Stéphan et al. 1976; Sparks et al. 1978; Herrmann 1979).

Radioactive halos, which are tiny zones of discolouration in certain minerals caused by the radiation from elements such as uranium and thorium, had been known since the early twentieth century. Gentry wrote several papers on the subject, claiming in 1968 to have discovered a new type of halo due to polonium (Gentry 1968). However, the existence of polonium halos was not confirmed and they remained controversial. Only in 1976 did Gentry suggest that anomalous radioactive halos constituted possible evidence for primordial SHEs. He presented his case at the 1978 Lubbock symposium on SHEs, but this time much more cautiously (Gentry 1978). The motivation for these papers was apparently scientific, but to Gentry, who was or developed into a young Earth creationist, they also had a different significance. He thought that his work on polonium halos in particular showed that the Earth was created divinely and in strict accordance with Genesis. As he phrased it in a book of 1986, the radioactive halos were "God's fingerprints in Earth's primordial rocks" (Numbers 2006, pp. 280–282).[2] In the early 1980s, after he had come out as a creationist, his association with the Oak Ridge laboratory was terminated and his scientific reputation declined rapidly.

Other scientists examined the possibility of stellar nucleosynthesis of SHEs. Given that absorption lines of technetium had been observed in the spectra of giant stars since 1952, it was not unreasonable to assume that also SHEs might be

---

[2]According to http://www.halos.com/, polonium halos prove that "our earth was founded in a very short time, in complete harmony with the Biblical record of creation." In the 1990s Gentry rejected the standard big bang cosmological model, arguing that the universe was static and divinely created.

produced by stellar nucleosynthesis (Merrill 1952). According to the prominent American astrophysicists David Schramm and William Fowler (1971), SHEs might indeed be produced in explosive stellar events. Their suggestion was found to be interesting but inconclusive as other calculations led to more pessimistic results. Suggestions of SHEs originating in neutron stars also did not lead to anything useful.

The search for natural SHEs was pursued with no less vigour in the Soviet Union than in the United States. One hunting ground was primary cosmic rays where heavy nuclear components had been found. Might there not only be heavy nuclei hidden in the rays but also superheavy nuclei? To answer the question, Russian physicists conducted elaborate experiments in space with the new artificial satellites. Neither these nor other searches revealed solid evidence for SHEs in cosmic rays (Zhdanov 1974).

During the 1970s Flerov and his Dubna co-workers conducted several searches of SHEs in meteorites, terrestrial minerals, geothermal waters and other natural sites. For a time the group supposed to have identified spontaneous fission events caused by a natural SHE, although they could not tell which. Closer examination and more precise measurements revealed that the evidence was far from convincing. All the same, the Dubna scientists engaged in new research projects aiming to detect SHEs in nature. At an international conference on the synthesis and properties of new elements held in Dubna in September 1980, Flerov and his colleague Gurgen Ter-Akopian admitted that so far no solid evidence for natural SHEs had been found. Despite the disappointing record they remained optimistic (Flerov and Ter-Akopian 1981, p. 921):

> Many chemists and physicists have been dreaming for years of revealing in nature chemical elements heavier than uranium. Rather often this dream seemed to be realizable and promised to open new horizons in the Sea of unstable heavy elements. Still more often it brought the disappointment of the hopes that did not come true. Nevertheless it is difficult for us to abandon this dream.

Since no convincing evidence of naturally occurring SHEs was found, latest by the mid-1980s the enthusiasm for further searches ceased. While unwilling to abandon the dream, Flerov and Ter-Akopian (1985, p. 380) were forced to agree with "the rather general view that the search for SHE in nature is hopeless." Superheavy elements seemed to be the exclusive business of nuclear laboratories. Claims of having detected the elements in ordinary matter were met with more than just ordinary scepticism, and yet scientists continued pursuing the search for SHEs in nature with still more advanced instruments (Dellinger et al. 2011; see also Sect. 4.1). The search is still going on. As late as 2015 Ter-Akopian essentially repeated what he had said 35 years earlier: "There is little chance that superheavy elements of no less than 100 million years are present on the stability island discovered at present. ... Nevertheless, the search for these nuclei in nature is justified in view of the fundamental importance of this topic" (Ter-Akopian and Dmitriev 2015, p. 177).

2.4   Motivations

Most scientists involved in SHE research consider it a field of fundamental science unrelated to applications. It is science for the sake of science, an adventure into the territory of still unknown nuclear physics. In a recent interview Yuri Oganessian justified his research field by saying that "it is about tackling fundamental questions in atomic physics." Foremost among the questions is the prediction of an island of stability. According to Oganessian, in an interview in Gray (2017): "Theorists predict that there should be some superheavy atoms, with certain combinations of protons and neutrons, that are extremely stable." Referring to elements heavier than $Z = 112$ he continued:

> Their lifetimes are extremely small, but if neutrons are added to the nuclei of these atoms, their lifetime grows. Adding eight neutrons to the heaviest known isotopes of elements 110, 111, 112 and even 113 increases their lifetime by around 100,000 times … but we are still far from the top of the island where atoms may have lifetimes of perhaps millions of years. We will need new machines to reach it.

Other leading SHE scientists have expressed a similar *l'art pour l'art* attitude. Sigurd Hofmann (2002, p. 205) refers to the "sense of the excitement which has motivated workers in this field" and suggests that the motivation for study SHEs is "because we are curious." Investigations of SHEs are about the limit of scientific knowledge, and that is what makes them exciting. According to Michael Nitschke, a member of Seaborg's and Ghiorso's group in Berkeley, "It is like not being sure whether or not the Earth is flat. You have to keep sailing. If you come to an edge and fall over, you form an entirely different conception of the universe than if you can just keep going. We are looking for the limits of matter—the edge of the world" (Browne 1978). In 1972 two nuclear chemists admitted that the primary motivation for SHE research was to "provide a testing ground for our understanding of the chemistry of the elements and the physics of the nucleus." But then, as an after-thought and referring to the possible discovery of long-lived SHEs in nature, they fell back on the mantra that "practical and useful applications would be forthcoming eventually, as is always the case with basic research" (Thompson and Tsang 1972, p. 1055). Yet another justification for early SHE research was that it might serendipitously lead to new discoveries, such as suggested by Keller at the 1978 Lubbock conference.

As has been the case with many earlier discoveries of elements, undoubtedly sheer pride and honour—including national and institutional honour—play an important role in the attempts to secure priority for a new SHE. Indeed, as will be abundantly clear in later chapters, the national element is obviously present in most of the controversies concerning element discoveries of this kind. As a British chemist engaged in SHE work wrote (Greenwood 1997, p. 179): "In a sense, it does not matter to science in general who actually discovers each new element. But it *does* matter to those who perhaps spend the bulk of their scientific careers on such projects. Personal pride, institutional pride and even national pride become involved and hundreds of millions of dollars are invested." He then added: "The main motivations are, of course, scientific: to gain a better understanding of nuclear structure and the forces which bind the subatomic particles; and to extend the chemical trends of the Periodic Table."

References

Atterling, H., et al.: Element 100 produced by means of cyclotron-accelerated oxygen ions. Phys. Rev. **95**, 585–586 (1954)

Barber, R.C., et al.: Discovery of the transfermium elements. Prog. Part. Nucl. Phys. **29**, 453–530 (1992)

Bhandari, N., et al.: Super-heavy elements in extraterrestrial samples. Nature **230**, 219–224 (1971)

Blokhintsev, D.I.: A decade of scientific work at the Joint Institute for Nuclear Research. Sov. Atom. Energy **20**, 328–345 (1966)

Bromley, D.A.: The development of heavy-ion physics. In: Bromley, D.A. (ed.) Treatise on Heavy-Ion Science, vol. 1, pp. 3–134. Plenum Press, New York (1984)

Browne, M.V.: Mentor and students test limits of matter. *New York Times*, 13 June (1978)

Dakowski, M.: The possibility of extinct superheavy elements occurring in meteorites. Earth Planet. Sci. Lett. **6**, 152–154 (1969)

Dellinger, F., et al.: Ultrasensitive search for long-lived superheavy nuclides in the mass-range $A = 288$ to $A = 300$ in natural Pt, Pb, and Bi. Phys. Rev. C **83**, 065806 (2011)

Donets, E.D., Shchegolev, V.A., Ermakov, V.A.: Synthesis of the isotope of element 103 (lawrencium) with mass number 256. Sov. Atom. Energy **19**, 995–999 (1965)

Flerov, G.N.: Synthesis and search for heavy transuranium elements. Sov. Atom. Energy **28**, 390–397 (1970)

Flerov, G.N.: History of the transfermium elements $Z = 101, 102, 103$. Sov. J. Part. Nucl. **22**, 453–483 (1991)

Flerov, G.N., Ter-Akopian, G.M.: The physical and chemical aspects of the search for superheavy elements. Pure Appl. Chem. **53**, 909–923 (1981)

Flerov, G.N., Ter-Akopian, G.M.: Superheavy elements. In: Bromley D.A. (ed.) Treatise on Heavy-Ion Science, vol. 4, pp. 333–402. Plenum Press, New York (1985)

Gareev, F.A., Kalinkin, B.N., Sobiczewski, A.: Closed shells for $Z > 82$ and $N > 126$ in a diffuse potential well. Phys. Lett. **22**, 500–502 (1966)

Gentry, R.V.: Fossil alpha-recoil analysis of certain variant radioactive halos. Science **160**, 1228–1230 (1968)

Gentry, R.V.: Are any unusual radiohalos evidence for SHE? Superheavy Elements. In: Lodhi, M. (ed.) Proceedings of the International Symposium on Superheavy Elements, pp. 123–152. Pergamon Press, New York (1978)

Gentry, R.V., et al.: Evidence for primordial superheavy elements. Phys. Rev. Lett. **37**, 11–15 (1976)

Ghiorso, A., et al.: New element, lawrencium, atomic number 103. Phys. Rev. Lett. **6**, 473–475 (1961)

Gray, R.: Breaking the periodic table. New Scientist 234 (15 April), 40–41 (2017)

Greenwood, N.N.: Recent developments concerning the discovery of elements 101–111. Pure Appl. Chem. **69**, 179–184 (1997)

Herrmann, G.: Superheavy-element research. Nature **280**, 543–549 (1979)

Herrmann, G.: Historical reminiscences: the pioneering years of superheavy element research. In: Schädel, M., Shaughnessy, D. (eds.) The Chemistry of Superheavy Elements, pp. 485–510. Springer, Berlin (2014)

Hodgson, P.: Discovery of superheavy elements. Nature **261**, 627 (1976)

Hoffman, D.C., et al.: Detection of plutonium-244 in nature. Nature **234**, 132–134 (1971)

Hoffman, D.C., Ghiorso, A., Seaborg, G.T.: The Transuranium People. Imperial College Press, London (2000)

Hofmann, S.: On Beyond Uranium: Journey to the End of the Periodic Table. Taylor and Francis, London (2002)

Jensen, K.A.: Nomenclature of inorganic chemistry. Pure Appl. Chem. **28**, 1–110 (1971)

Kalinkin, B.N., Gareev, F.A.: On the problem of synthesis of superheavy nuclei: A short historical review on first theoretical predictions and new experimental reality (2001). arxiv:nucl-ex/0105021

Keller, O.L.: History and perspective of the search for superheavy elements In: Lodhi, M. (ed.) Superheavy Elements Proceedings of the International Symposium on Superheavy Elements, pp. 10–21. Pergamon Press, New York (1978)

Kragh, H.: An unlikely connection: Geochemistry and nuclear structure. Phys. Perspect. **2**, 381–397 (2000)

Lodhi, M.A.K. (ed.): Superheavy elements. In: Proceedings of the International Symposium on Superheavy Elements. Pergamon Press, New York (1978)

Maglich, B. (ed.): Adventures in Experimental Physics, vol. 2. Princeton, World Science Education (1972)

Merrill, P.W.: Technetium in the stars. Science **115**, 484 (1952)

Mladenović, M.: The Defining Years of Nuclear Physics 1932–1960s. Institute of Physics Publishing, Bristol (1998)

Myers, W., Swiatecki, W.: Nuclear masses and deformations. Nucl. Phys. **81**, 1–60 (1966)

Nilsson, S.G., et al.: On the nuclear structure and stability of heavy and superheavy elements. Nucl. Phys. A **131**, 1–66 (1969)

Nilsson, S.G., Nilsson, N.R. (eds.): Super-heavy elements: Theoretical predictions and experimental generation. *Physica Scripta* **10A**, 1–186 (1974)

Numbers, R.L.: The Creationists: From Scientific Creationism to Intelligent Design. Harvard University Press, Cambridge, MA (2006)

Schramm, D.N., Fowler, W.A.: Synthesis of superheavy elements in the r-process. Nature **231**, 103–106 (1971)

Seaborg, G.T.: Terminology of the transuranium elements. Terminology. 1, 229-252 (1994)

Sparks, C.J., et al.: Evidence against superheavy elements in giant-halo inclusions re-examined with synchrotron radiation. Phys. Rev. Lett. **40**, 507–511 (1978)

Stéphan, C., et al.: Search for superheavy elements in monazite ore from Madagascar. Phys. Rev. Lett. **37**, 1534–1536 (1976)

Stoughton, R.W., et al.: A search for naturally occurring superheavy elements. Nature Phys. Sci. **246**, 26–28 (1973)

Stradins, J.P., Trifonow, D.N., Pijola, S.: Die Evolution der Idee von Inseln Relativer Stabilität der Chemischen Elemente. D.A.V.I.D. Verlagsgesellschaft, Berlin (1987)

Sullivan, W.: Superheavy element is believed found. New York Times, 18 June (1976)

Ter-Akopian, G.M., Dmitriev, S.N.: Searches for superheavy elements in nature: Cosmic-ray nuclei; spontaneous fission. Nucl. Phys. A **944**, 177–189 (2015)

Thompson, S.G., Tsang, C.F.: Superheavy elements. Science **178**, 1047–1055 (1972)

Tsaletka, R., Lapitskii, A.V.: Occurrence of the transuranium elements in nature. Russ. Chem. Rev. **79**, 684–689 (1960)

Wilkinson, D.H., et al.: Discovery of the transfermium elements. Pure Appl. Chem. **67**, 1757–1814 (1993)

Zhdanov, G.B.: Search for transuranium elements (methods, results, and prospects). Sov. Phys. Usp. **16**, 642–658 (1974)

Chapter 3
On Element Discoveries

Abstract While there is little ambiguity in the definition of a chemical element, it is far from evident what it means to have discovered a new element and what the criteria for discovery are. In the post-World War II era recognition of new elements was the responsibility of IUPAC, the International Union of Pure and Applied Chemistry, which also had the final decision regarding names. The synthesis of still more transuranic elements and the corresponding priority claims resulted in the mid-1980s in the so-called Transfermium Working Group (TWG) established jointly by IUPAC and the physicists' union IUPAP. However, the recommendations of TWG only created more controversies. Another and earlier discovery controversy concerned element 102, the priority of which was claimed by Swedish, American and Russian research groups. Despite an extended controversy, the originally proposed name, nobelium, was accepted by IUPAC.

Keywords Discovery · Chemical elements · Priority · Controversies
IUPAC · Transfermium working group · Nobelium

The discoveries of superheavy elements have in many cases been controversial, in part because different groups of researchers have claimed priority to the discovery of the same element. Some of the controversies were fuelled by the political tensions of the Cold War period. But what are, more precisely, the criteria for having discovered a new element and who have the authority to formulate the criteria and validate discovery claims? In addition to discussing these questions in their historical contexts, this chapter also analyses in some detail an early controversial discovery story which only came to a close after more than thirty years. The element in question is nobelium, element number 102 in the periodic table. The general account of discovery criteria and the naming of elements will be followed up in Chap. 5.

© The Author(s), under exclusive licence to Springer International Publishing AG, part of Springer Nature 2018
H. Kragh, *From Transuranic to Superheavy Elements*, SpringerBriefs in History of Science and Technology, https://doi.org/10.1007/978-3-319-75813-8_3

3.1   General Considerations

Many of the discoveries of transuranic elements and the naming of them have been
surrounded by controversies of priority often involving national and political elements. This is hardly surprising, given that as a rule discoveries of new elements
occurring in nature have also been followed by controversies of this kind. Early
examples are oxygen and vanadium, and later examples are hafnium (Kragh 1980)
and rhenium (Zingales 2005; Scerri 2013). While most of the priority conflicts have
been contemporaneous with the discovery claims, in some cases they have occurred
retrospectively and been due to later intervention of scientists or historians of
science. But of course priority disputes are not limited to chemical elements as they
constitute a general feature in science. As pointed out by the sociologist Robert
Merton in a classical study first published 1957, priority conflicts are part and parcel
of the development of science and not anomalous features (Merton 1973; Gross
1988). Moreover, they play a significant and often progressive role by forcing the
participants in a priority conflict to study more extensively the issue under consideration than if the conflict had not existed.

Naturally, a priority conflict concerning a new element can only be resolved on
basis of some commonly agreed criteria of what constitutes a discovery. According
to Rancke-Madsen (1976), two conditions are necessary and sufficient for a person
to be accepted as the discoverer of an element:

> 1. He has observed the existence of a new substance which is different from earlier
> described substances, and this new substance is recognized by him or later by scientists as
> being elemental. 2. He must have published the discovery of the new substance in such a
> manner that it has been noticed by contemporaries outside the immediate circle of the
> discoverer, usually in a periodical or monograph.

These conditions are reasonable, but they fail to discriminate between discovery
claims and discoveries, or between claimed observations and scientifically valid
observations. Moreover, they do not take into account that priority disputes are
settled by negotiations according to social norms within the relevant scientific
community and that these norms may change over time. On whose authority are the
norms and criteria based? What if individual scientists do not accept the criteria
shared by the scientific community at large?

Since the early twentieth century the recognition and naming of new elements
have been formalized and institutionalized. In the earlier period the unwritten rule
was that the discoverer named the element, although there was no agreed system
and in many cases also no agreement as to who the discovery should be credited.
The naming of elements through history is surveyed in Childs (1998).

The first national atomic weight committees appeared in the 1890s, starting with
a committee appointed by the American Chemical Society (ACS) in 1892.
Remarkably, this committee consisted of just a single chemist, Frank W. Clarke
from the U.S. Geological Survey. Together with the ACS one-man committee, the
most important of the national committees was the German Atomic Weight
Commission (Deutsche Atomgewichtskommission) formed in 1897. In 1899 the

International Committee on Atomic Weights (ICAW) was established, consisting of no less than 56 delegates from eleven countries (Crawford 1992, p. 40; Holden 2004). The committee was soon slimmed to the more manageable size of three members, one from the United States, one from Germany and one from England. The early tables of atomic weight were based on the O = 16 scale rather than the H = 1 scale. Only in 1961 did the chemical community officially adopt the new scale defined by C-12 = 12 which approximately agrees with the H = 1 scale going back to Dalton in the early nineteenth century.

An International Association of Chemical Societies was formed in Paris in 1911, but it lasted only to the beginning of World War I. After the war, ICAW was reorganized within the structure of IUPAC which was founded in 1919 as part of IRC, the International Research Council. However, initially only the chemical societies from the allied powers and neutral countries were admitted to IUPAC. In connection with the German membership in 1930 IUPAC was renamed IUC (International Union of Chemistry), but in 1947 the organization returned to its earlier name (Fennell 1994). Much later, in 1979, the atomic weight committee's name was changed to the Commission on Atomic Weights and Isotopic Abundances which again, in 2002, was changed to the Commission on Isotopic Abundances and Atomic Weights, CIAAW.

The new definition of an element in terms of its atomic number instead of its weight was adopted by the German Atomic Weight Commission in 1921 and two years later also by a new committee under IUPAC, the International Committee on Chemical Elements. However, not all chemists and especially not those of the older generation were happy with the physics-based definition, which they felt was foreign to the chemical mind (Kragh 2000). How could one call a non-elemental substance an element, they asked? Much later IUPAC faced a minor controversy regarding the meaning of "atomic weight" in relation to the concept of atomic mass, isotopes of non-natural elements, and other issues (de Biévre and Peiser 1992). The compromise result was that "An atomic weight (relative atomic mass) of an element from a specified source is the ratio of the average mass per atom of the element to 1/12 of the mass of an atom of ^{12}C."

In the early days the names of new elements were not officially approved by ICAW except that they entered the committee's versions of the periodic table and in this sense received the blessing of IUPAC. In the late 1940s questions of names and priority were conferred to IUPAC's Commission on Nomenclature of Inorganic Chemistry (CNIC) which since then has played an important if sometimes contested role relating to SHE discoveries. CNIC was terminated in 2002 and its responsibilities transferred to the Inorganic Chemistry Division under IUPAC.

At an IUC-IUPAC conference in London in 1947 it was agreed that only CNIC could recommend a name to the chemical union and that the final decision remained with IUPAC's Executive Council. "It has been accepted in the past that the discoverers of a new element had the sole right to name it," it was stated (Koppenol 2002, p. 789). But IUPAC did not agree with this historical tradition:

Table 3.1 Elements recognised by IUPAC in 1949

Z	Name	Symbol	Country	Year	Discoverers
43	technetium	Tc	Italy	1937	E. Segré, C. Perrier
61	promethium	Pm	USA	1945	J. Marinsky, E. Glendenin, C. Coryell
85	astatine	At	USA	1940	D. Corson, K. MacKenzie, E. Segré
87	francium	Fr	France	1939	M. Perey
93	neptunium	Np	USA	1940	E. McMillan, P. Abelson
94	plutonium	Pu	USA	1941	G. Seaborg et al.
95	americium	Am	USA	1944	G. Seaborg et al.
96	curium	Cm	USA	1944	G. Seaborg, R. James, A. Ghiorso

Priority is only one factor to be considered in deciding which is the best name for general international adoption. This presumptive right to name new elements is now accorded to the discoverers of new elements produced artificially, but subject to the approval of the Nomenclature Commission of IUPAC.

The names of the first transuranic elements were considered at the 15th IUC-IUPAC conference in 1949, taking place in Amsterdam at a time when the International Union embraced 31 countries. At this occasion CNIC officially adopted the proposed names for elements 93 to 96 and also decided upon the names of some other elements (Anon. 1949; Coryell and Sugarman 1950; Koppenol 2005). For example, it was on this occasion that element 4 officially became beryllium and the older name glucinium was ruled out. It was also on this occasion that the name astatine was adopted for element 85, technetium for element 43 at the expense of masurium and some other names, and niobium for element 41 instead of columbium. Element 74 was now officially called wolfram, whereas the name tungsten was abandoned if only to be reinstated with the same symbol W later on. Names and symbols of the eight new elements recognised at the Amsterdam conference are given in Table 3.1.

Contrary to the later troubles with assigning places and names for the synthetic elements, the IUC decision was straightforward: "Elements 93, 94, 43, and 85 were discovered at the University of California; 95 and 96 were discovered at the Metallurgical Laboratory, Chicago; and 61 was discovered at Oak Ridge" (Anon. 1949, p. 2996).

During the first decades of the twentieth century there were two basic criteria for recognizing the discovery of a new element, namely the optical spectrum and the atomic weight of the claimed element (Kragh 2000). At an international conference on applied chemistry held in Paris in 1900 it was decided that a new element should only be accepted if it was prepared in such quantity that its atomic weight could be determined and its spark spectrum be analysed and shown to differ from all other elements (Holden 2004). On the other hand, recognition of the existence of a new element did not necessarily require that it was prepared in a pure state. One example is radium and an earlier one is fluorine, which was first isolated by Henri Moissan in 1886.

Since the mid-1920s the main criterion became the element's characteristic X-ray spectrum, which was first used for the discovery of hafnium in 1923 and a few years later also for rhenium. While SHEs have no definite atomic weight, in many cases they can and have been identified by means of the X-ray method. "The best evidence for the identification of a new element," wrote a group of American scientists, "is the observation of characteristic K- or L-series X rays, as first utilized by Moseley in 1914" (Bemis et al. 1973). However, in some other cases SHEs have no spectrum based on electron transitions between different atomic energy levels. Besides, X-ray spectroscopy requires amounts of matter much greater than the few atoms often produced in SHE reactions. When the method could nonetheless be used, it was often indirectly, namely by determining the characteristic X-rays emitted by atoms of the daughter nuclide after the decay of the parent SHE nuclide.

In 1978, after much delay and many years of preparation, CNIC suggested a naming procedure based directly on the atomic number and aimed specifically for the elements with $Z > 103$ which had not yet been discovered. The provisional names and corresponding symbols were primarily based on Latin numbers, for instance unnilpentium (un-nil-pentium, Unp) for $Z = 105$ and ununbium (un-un-bium, Uub) for $Z = 112$. However, in some cases the names included the Greek prefixes penta and hexa, as in elements 105 and 106. There was room even for $Z = 999$, which would be enenenium (en-en-enium). Contrary to the symbols of the known elements, the systematic symbols consisted of three letters. According to CNIC, the systematic names did not affect the already recommended names for elements 101 to 103. Moreover, "The systematic names and symbols for elements of atomic numbers greater than 103 are the only approved names and symbols for those elements until the approval of trivial names by IUPAC" (Chatt 1979; see also Orna 1982 and Bera 1999).

But although systematic and neutral, the unwieldy names were rarely used by the scientists who preferred to refer directly to the atomic numbers. As a linguist noted, the system was "a masterpiece of diplomacy but also of blandness and lack of imagination, as well as an etymological hodge-podge" (Diament 1991, p. 210). A search in Google Scholar shows that in the period 1980–1995 not a single research paper referred to unnilhexium, the systematic name for element 106 later to be named seaborgium. More than 1,000 papers referred to "element 106" and about 70 to "seaborgium."

The question of the SHEs came up during the 33rd IUPAC General Assembly held in Lyon in the summer of 1985. On this occasion the chairman of the Commission on Atomic Weights, Norman Holden from the Oak Ridge National Laboratory, presented a detailed status report on the discovery histories of elements from $Z = 104$ to 109 (Holden 1985). To take care of priority questions—a subject which Holden had avoided—IUPAC and IUPAP established the same year the Transfermium Working Group (TWG) consisting of nine scientists, two of which were nominated by IUPAC and seven by IUPAP. The uneven distribution possibly reflects that the initiative came from IUPAP's president Allan Bromley, a Canadian-American nuclear physicist and science administrator, although it was originally on the instigation of IUPAC (Barber et al. 1992, p. 455). According to Bromley, this is what happened:

> When I was president of IUPAP I was contacted by the then president of IUPAC, then the director of the chemistry division of the National Research Council in Ottawa, who told me about the efforts that IUPAC had been making to get some kind of consensus concerning the naming of the transuranics generally. He admitted that they had had repeated failures and asked for IUPAP's help. To that end I appointed a committee chaired by Sir Denys Wilkinson that took their task very seriously.[1]

Initially the working group consisted solely of physicists and it was only after objections from the chemistry community that it was extended with the two IUPAC representatives, Norman Greenwood from England and Yves Jeannin from France. According to Paul Karol (2004), who at the time was a member of the IUPAC Commission on Radiochemistry and Nuclear Techniques and later served as its Chair, the two chemists were appointed without conferring with the commission. The members appointed by IUPAP were from England (D. Wilkinson), Canada (R. Barber), Poland (A. Hyrnkiewicz), France (M. Lefort), Japan (M. Sakai), Czechoslovakia (I. Ulehla), and The Netherlands (A. Wapstra).

Scientists from USA, USSR and West Germany were deliberately excluded from TWG to keep the group free of national bias, but of course the group had consultations with SHE scientists from these countries. In December 1988 the TWG visited the laboratory in Darmstadt, Germany (see also Sect. 4.1), after which it had a meeting in Berkeley in June 1989. The two meetings are described in Armbruster and Münzenberg (2012, pp. 283–284) and in Hoffman et al. (2000, pp. 371–378). A planned TWG visit to Dubna was delayed until February 1990. In addition to these visits, the TWG also contacted other scientists involved in SHE synthesis. The responsibility of the group, headed by the distinguished British nuclear physicist Denys Wilkinson, was to formulate criteria for when an element was discovered and to evaluate discovery claims accordingly.[2] On the other hand, the naming of new elements was not part of the group's agenda. Wilkinson emphasised the neutrality and objectivity of the working group (TWG 1993, p. 1824):

> The TWG consisted of nine men of goodwill who, conjointly and severally, spent some thousands of hours ... in a microscopic and scrupulous analysis of the discovery of the transfermium elements. ... We were utterly without bias, prejudice or pre-commitment and had no connection with any of the laboratories of chief concern; we did not care who had discovered the elements in question but agreed to find out.

---

[1]Bromley to Krishna Kumar, 5 January 1998. Facsimile reproduction in http://www.marinov-she-research.com/image/users/352030/ftp/my_files/External/Bromley%20Letter%201998.pdf.
The IUPAC president was William G. Schneider, who served in the position 1983–1985.

[2]Denys H. Wilkinson (1922–2016) served from 1962 to 1976 as head of the Department of Nuclear Physics, Oxford University, and subsequently as Vice-Chancellor of the University of Sussex. A member of the Royal Society since 1956, he was knighted in 1974. Fontani, Costa and Orna (2015), p. 386 states mistakenly that the TWG chairman was Geoffrey Wilkinson (1921–1996), the 1973 Nobel Prize laureate in chemistry.

TWG held its first meeting in February 1988 and published its criteria three years later in *Pure and Applied Chemistry*, since 1960 the official journal of IUPAC (Wapstra 1991). In 1991 the TWG was disbanded and eight years later it was followed by another inter-union group of experts, called the Joint Working Party (JWP). This party or group consisted of only four members (an American, a Briton, a Japanese and a Canadian), with two representing each of the participating scientific unions. The Chair of this group was the American nuclear chemist Paul Karol, an indication that the former exclusion of members from claimant countries was no longer maintained. Later again a new five-member joint working group, this time with the Canadian physicist Robert Barber as Chair, was established in 2011 to examine claims of having discovered elements with $Z > 112$.

3.2 Nobelium, A Controversial Element

The first two decades of manufacturing and analysing new transuranic elements were dominated by Seaborg and his group of Californian physicists and nuclear chemists. Their dominance was first challenged in connection with element 102, which was announced in 1957 by a group of Swedish, American and British physicists working at Stockholm's Nobel Institute of Physics. A team of Swedish physicists had three years earlier used the institute's 225 cm cyclotron to bombard a uranium target with O-16 ions. From a measured alpha activity of energy 7.7 meV and half-life approximately half an hour they suggested to have produced an isotope of $Z = 100$ but without determining its mass number except that it might be $A \cong 250$ (Atterling et al. 1954; see also Sect. 2.1). In the 1957 experiment the extended group led a beam of C-13 ions to react with a sample of curium containing 94% Cm-244. The group based its discovery claim on the measured activity of an alpha emitter with energy 8.5 ± 0.1 meV and half-life of about 10 min. They suggested that the mass number of the nuclide of element 102 was either 251 or 253, produced by

$$^{13}_{6}C + ^{244}_{96}Cm \rightarrow ^{251}_{102}Z + 6^{1}_{0}n \quad \text{or} \quad ^{13}_{6}C + ^{244}_{96}Cm \rightarrow ^{253}_{102}Z + 4^{1}_{0}n$$

The Stockholm physicists confidently announced not only a new transuranic element but also suggested "the name nobelium, symbol No, for the new element in recognition of Alfred Nobel's support of scientific research and after the institute where the work was done" (Fields et al. 1957). The discovery claim created much attention and was covered not only in *Svenska Dagbladet*, a leading Swedish newspaper, but also in *The Guardian* and other British news media. The public interest was in part scientifically motivated, but there were other reasons as well. As John Milsted, one of the British members of the Stockholm team noted, "This is the first of the trans-uranium elements to be discovered on European soil, and the first to be found by an international effort" (Milsted 1957; see also McKay 1957). In accordance with the climate of the Cold War, apparently Milsted did not admit Dubna to be part of Europe!

In England, communications on the new element were presented at the meeting of the British Association for the Advancement of Science in Dublin in September 1957. A little earlier John Maddox, a physics-trained science writer and later the editor of *Nature*, introduced nobelium to readers of *The New Scientist*. Maddox (1957) described the discovery as "in every way comparable with the discovery of a new mountain peak or some old buried city." Given that the highly unstable nobelium atoms were created in the laboratory, the comparison was not very fortunate. Nor was Maddox's suggestion that "element 102 may turn out to be the tenth and last of synthetic elements" a fortunate one.

The Swedish-British-American discovery claim turned out to be unfounded or at least insufficient. It gave rise to other and rival discovery claims by American and Russian teams which made nobelium a controversial element as far as priority was concerned. Karol (2004, p. 256) called its discovery "the most convoluted and misunderstood of all the transfermiums." Accounts of the nobelium controversy, as seen from an American perspective, include Ghiorso and Sikkeland (1967) and Seaborg (1968, pp. 61–65). For the Russian perspective, see Flerov et al. (1992).

The Berkeley team initially set out to confirm the findings of the Swedish-British group, assuming that "they were correct since the Stockholm team had published its work in a well-written paper in the peer-reviewed, highly respected journal *The Physical Review*" (Hoffman, Ghiorso and Seaborg 2000, p. 234). But it soon turned out that the results obtained in Stockholm could not be reproduced by the Berkeley scientists. Reportedly, the Berkeley group wondered in private if "nobelievium" would be a more appropriate name (Thornton and Burdette 2014). In experiments of 1958 Ghiorso, Seaborg, Torbjorn Sikkeland and John Walton announced to have positively identified the isotope No-254 and thus to have discovered the element (Ghiorso et al. 1958). A target of curium was bombarded with carbon-12 ions,

$$^{12}_{6}C + {}^{246}_{96}Cm \rightarrow {}^{254}_{102}Z + 4{}^{1}_{0}n$$

The suspected nobelium isotope (Z) was identified by its alpha decay and the known radioactive properties of the fermium daughter nucleus:

$$^{254}_{102}Z \rightarrow {}^{250}_{100}Fm + {}^{4}_{2}He$$

The Americans were confident that they had discovered the element and that the 1957 discovery claim was unfounded. The Stockholm group admitted that the Berkeley experiments "appeared to cast some doubt on our results" but concluded in a re-examination of early 1959 that the doubt was apparent only. Although the group maintained the validity of the 1957 investigation, except that they now found the production of nuclide 255 from Cm-246 to be more likely, "we suggest that judgment on the discovery of element 102 should be reserved" (Fields et al. 1959). Here is how *Britannica Book of the Year 1959* presented the situation (Armitage 1959, p. 129):

In 1958 some doubt was thrown on previous year's report that an isotope of element 102 with a half-life of ten minutes had been discovered by a team of U.S., British and Swedish scientists who had named it Nobelium. Workers at Berkeley, where most of the transuranic elements had been first synthesized, failed to repeat the synthesis of element 102 by bombarding curium-244 with carbon-13 nuclei. ... On the other hand workers in the U.S.S.R. who bombarded plutonium-241 with oxygen-16 nuclei appeared to confirm the earlier findings.

However, the results reported by the Stockholm group could also not be reproduced by Georgii Flerov and his collaborators at the Kurchatov Institute in Moscow, the predecessor of the Joint Institute for Nuclear Research (JINR) in Dubna. Flerov was a highly decorated Soviet nuclear physicist who back in 1940, together with his colleague Konstantin Petrzhak, had discovered spontaneous fission in U-238 and a few years later got involved in the Soviet atomic bomb project. Not only did the Russian physicists disbelieve the news from Stockholm, they also considered the American experiments to be no more than just an indication of $Z = 102$. Although they did not claim to have discovered element 102 in their early experiments from 1957 and 1958, in a series of later experiments lasting until 1966 they provided definite proof of the existence of isotopes of the element. The Russians pointed to errors and inadequacies in the American experiments and they obtained results that flatly contradicted Ghiorso's 1958 claim of having discovered the $A = 254$ isotope.

As Flerov and his Dubna group saw it, there was no doubt where the priority belonged: "Element 102 was discovered at Dubna in studies carried out during 1963–1966. Those papers contain unambiguous and complete evidence for the synthesis and observation of its nuclei, as well as correct data on the α-decay properties of five isotopes of element 102" (Flerov et al. 1992, p. 123; Flerov 1970). On the other hand, although the Americans admitted to have been wrong on some points they disbelieved some of the Russian results and suggested that they did not justify priority to the discovery of the new element. Ghiorso and Seaborg (1992) immediately responded to what they considered the selective and misrepresented history of element 102 presented by Flerov and his co-authors. If the element had really been identified by Soviet discoverers, who were they? "Perhaps," the two Berkeley scientists asked, "the Dubna group should nominate their candidates to be identified as co-discoverers of this element." Clearly, remains of the Cold War atmosphere of disbelief between the two superpowers contributed to the controversy between the two groups.

As to the name of element 102, although the Dubna researchers refrained from using "nobelium" they did not suggest a different name. At first the Americans wanted to replace what they considered the too hastily assigned name nobelium with another one which better reflected the actual discovery. "Although the name nobelium for element 102 will undoubtedly have to be changed," wrote Seaborg (1959, p. 11; 1963, p. 33), "the investigators have not, at the time of writing, made their suggestion for the new name." However, at the 21st IUPAC conference in 1961 the name and symbol received approval by entering the ICAW list of chemical elements (Cameron and Wichers 1962). Reflecting the uncertainty at the time, no isotope or mass number was associated with the new element. Apart from being welcomed by IUPAC, nobelium had also come into common usage and entered textbooks and periodic tables. For these and other reasons the Americans refrained from proposing an alternative name. The Russian group, systematically avoiding the name nobelium, asserted its priority and right to suggest a name. "Joliotium" with chemical symbol Jl, referring to the French physicist, Nobel laureate and devoted communist Frédéric Joliot, was for a time the preferred name in the Soviet Union and later proposed also for the elements of atomic numbers 103 and 105. Joliotium was recommended for element 105 by IUPAC in 1994, but replaced by dubnium three years later (Fontani, Costa and Orna 2015, pp. 385–388; Rayner-Canham and Zheng 2008).

It took decades before the conflict over name and priority of the new element was finally settled. Only in reports from the early 1990s did IUPAC announce its decision. After a review of all relevant papers 1957–1971 the TWG concluded in favour of Dubna experiments of 1966 which "give conclusive evidence that element 102 had been produced" (Barber et al. 1992). The TWG failed to find the same kind of evidence in the Berkeley work of 1958 which consequently was not accepted as the discovery of the element. The Berkeley group strongly disagreed, arguing that at least it should share the credit with the Russians. Seaborg (1994, p. 242) found the TWG decision to be "incredible" and "erroneous."

The American specialists in SHE research protested vehemently and in general terms to the recommendations of TWG concerning the elements of atomic numbers 101–112. Ghiorso and Seaborg (1993) plainly charged that the TWG panel was incompetent and their report "riddled with errors of omission and commission." Among the complaints was that the panel, consisting of seven IUPAP members and only two IUPAC members, was biased toward physics with the consequence that the report was marked by a "downgrading of chemical contributions." Although not mentioning Wilkinson by name, Ghiorso and Seaborg made it clear that he was primarily to be blamed for the report's lack of appreciation of nuclear-chemical methods. Seaborg was trained in chemistry and considered himself a nuclear chemist and not a nuclear physicist. Since the Berkeley group relied much on expertise in nuclear chemistry, it thought that the alleged physics bias led to an unfair evaluation of the group's work. Paul Karol, a nuclear chemist at the Carnegie Mellon University, agreed. Calling the TWG study "flawed," he deplored that it included no experts in SHE chemistry (Karol 1994). The only TWG member with experience in SHE research was the French physicist Marc Lefort from the Orsay Institute of Nuclear Physics in northern France. The German attitude to the TWG was much more positive than the one of their American colleagues (Armbruster and Münzenberg 2012, p. 284).

Moreover, Ghiorso and Seaborg complained that the TWG had had meetings with the Dubna scientists, but none with the people from Berkeley, prior to the publication of the report. What Ghiorso and Seaborg did not comment on, but undoubtedly had in mind, was the national composition of the TWG. The nine members included one from Japan and eight Europeans, two of them from the former Soviet bloc. There were no Americans. Although the Soviet Union was dissolved by the end of 1991, the Cold War was not a forgotten chapter.

References

Anon.: Names of new elements confirmed by International Union of Chemistry. Chem. Eng. News **27**, 2996–2999, 3093 (1949)

Armbruster, P., Münzenberg, G.: An experimental paradigm opening up the world of superheavy elements. Eur. Phys. J. H **37**, 327–410 (2012)

Armitage, J. (ed.): Britannica Book of the Year 1959. Encyclopedia Britannica, London (1959)

Atterling, H., et al.: Element 100 produced by means of cyclotron-accelerated oxygen ions. Phys. Rev. **95**, 585–586 (1954)

Barber, R.C., et al.: Discovery of the transfermium elements. Prog. Part. Nucl. Phys. **29**, 453–530 (1992)

Bemis, C.E., et al.: X-ray identification of element 104. Phys. Rev. Lett. **31**, 647–649 (1973)

Bera, J.K.: Names of the heavier elements. Resonance 4(March), 53–61 (1999)

Cameron, A.E., Wichers, E.: Report on the international commission on atomic weights. J. Am. Chem. Soc. **84**, 4174–4197 (1962)

Chatt, J.: Recommendations for the naming of atomic numbers greater than 100. Pure Appl. Chem. **51**, 381–384 (1979)

Childs, P.E.: From hydrogen to meitnerium: naming the chemical elements. In: Thurlow, K. (ed.) Chemical Nomenclature, pp. 27–66. Kluwer Academic, Dordrect (1998)

Coryell, C.D., Sugarman, N.: The acceptance of new official names for the elements. J. Chem. Educ. **27**, 460–461 (1950)

Crawford, E.: Nationalism and Internationalism in Science, 1880–1939. Cambridge University Press, Cambridge (1992)

De Bièvre, P., Peiser, H.S.: 'Atomic weight'—the name, its history, definition, and units. Pure Appl. Chem. **64**, 1535–1543 (1992)

Diament, H.: Politics and nationalism in the naming of chemical elements. Names A J. Onomast. **39**, 203–216 (1991)

Fennell, R.: History of IUPAC 1919–1987. Blackwell Science, London (1994)

Fields, P.R., et al.: Production of the new element 102. Phys. Rev. **107**, 1460–1462 (1957)

Fields, P.R., et al.: On the production of element 102. Arkiv för Fysik **15**, 225–228 (1959)

Flerov, G.N.: Synthesis and search for heavy transuranium elements. Sov. Atom. Energy **28**, 390–397 (1970)

Flerov, G.N., et al.: A history and analysis of the discovery of element 102. Radiochim. Acta **56**, 111–124 (1992)

Fontani, M., Costa, M., Orna, M.V.: The Lost Elements: The Periodic Table's Shadow Side. Oxford University Press, Oxford (2015)

Ghiorso, A., et al.: Element no. 102. Phys. Rev. Lett. **1**, 18–21 (1958)

Ghiorso, A., Seaborg, G.T.: Comments on the selective history and analysis of the discovery of element 102. Radiochim. Acta **56**, 125–126 (1992)

Ghiorso, A., Seaborg, G.T.: Response to the report of the transfermium working group discovery of the transfermium elements. Prog. Part. Nucl. Phys. **31**, 233–237 (1993)

Ghiorso, A., Sikkeland, T.: The search for element 102. Phys. Today **20**, 25–32 (1967)

Gross, A.G.: Do disputes over priority tell us anything about science? Sci. Context **11**, 161–179 (1988)

Hoffman, D.C., Ghiorso, A., Seaborg, G.T.: Transuranium People: The Inside Story. Imperial College Press, London (2000)

Holden, N.E.: Status of the translawrencium elements. Unpublished memorandum (1985). https://www.osti.gov/scitech/servlets/purl/5112603

Holden, N.E.: Atomic weights and the international committee–a historical review. Chem. Int. **26** (1) (2004). https://www.iupac.org/publications/ci/2004/2601/1_holden.html

Karol, P.J.: Transfermium wars. Chem. Eng. News **18**, 2–3 (1994)

Karol, P.J.: The heavy elements. In: Ouvray D.H., King, R.B. (eds.) The Periodic Table: Into the 21st Century, pp. 235–262. Research Studies Press, Baldock, UK (2004)

Koppenol, W.H.: Naming of new elements. Pure Appl. Chem. **74**, 787–791 (2002)

Koppenol, W.H.: Paneth, IUPAC, and the naming of elements. Helv. Chim. Acta **88**, 95–99 (2005)

Kragh, H.: Anatomy of a priority conflict: the case of element 72. Centaurus **23**, 275–301 (1980)

Kragh, H.: Conceptual changes in chemistry: the notion of a chemical element. Stud. Hist. Philos. Mod. Phys. **31**, 435–450 (2000)

Maddox, J.: Element 102. New Sci. **2**, 26–28 (1957)

McKay, H.A.C.: Preparation of trans-uranium elements. Nature **180**, 1010–1012 (1957)

Merton, R.K.: Priorities in scientific discovery. In: Merton, R.K. (ed.) The Sociology of Science: Theoretical and Empirical Investigations, pp. 286–324. University of Chicago Press, Chicago (1973)

Milsted, J.: The preparation of nobelium. Nature **180**, 1012–1013 (1957)

Orna, M.V.: On naming the elements with atomic number greater than 100. J. Chem. Educ. **59**, 123 (1982)

Rancke-Madsen, E.: The discovery of an element. Centaurus **19**, 299–313 (1976)

Rayner-Canham, G., Zheng, Z.: Naming elements after scientists: an account of a controversy. Found. Chem. **10**, 13–18 (2008)

Scerri, E.: A Tale of 7 Elements. Oxford University Press, Oxford (2013)

Seaborg, G.T.: The transuranium elements. Endeavour **18**, 5–13 (1959)

Seaborg, G.T.: Man-Made Transuranium Elements. Prentice-Hall, Englewood Cliffs, N. J (1963)

Seaborg, G.T.: Elements beyond 100, present status and future prospects. Annu. Rev. Nucl. Part. Sci. **18**, 53–152 (1968)

Seaborg, G.T.: Terminology of the transuranium elements. Terminology **1**, 229–252 (1994)

Thornton, B.F., Burdette, S.C.: Nobelium non-believers. Nature Chem. **6**, 652 (2014)

TWG: Responses on the report discovery of the transfermium elements. Pure Appl. Chem. **65**, 1815–1824 (1993)

Wapstra, A.H.: Criteria that must be satisfied for the discovery of a new chemical element to be recognized. Pure Appl. Chem. **63**, 879–886 (1991)

Zingales, R.: From masurium to trinacrium: the troubled story of Element 43. J. Chem. Educ. **82**, 221–227 (2005)

Chapter 4
Failed Discovery Claims

Abstract Without suggesting a name, in 1971 Amnon Marinov and collaborators announced to have detected element 112 by bombarding a tungsten target with high-energy protons. The discovery claim was not accepted by specialists in the synthesis of superheavy elements who were unable to replicate the experiment. Nonetheless, Marinov and members of his group insisted that they had discovered the element. They later claimed to have found long-lived nuclides of element 122 in natural samples, a claim which was also not accepted. The other case described in the chapter is quite different in substance but also concerns a failed discovery claim. The announcement in 1999 that a research group in Berkeley had produced element 118 turned out to be based on false data, namely fraud committed by Viktor Ninov, a member of the group. The Ninov affair caused much alarm, not only in the scientific community but also generally. Element 118 was eventually synthesised, but only in 2016 was its existence officially approved.

Keywords Superheavy elements · Amnon Marinov · Viktor Ninov
Scientific misconduct · Element 112 · Element 118

The history of science is full of discovery claims which have not stood the test of time in the sense that they were at some stage shown to be wrong. What the scientist thought to have discovered did not correspond to later knowledge of what really exist. Such failed discovery claims are part and parcel of the development of science and although wrong, they often play a positive role in the development of science. The difference between true and false discoveries, or between those which are successful and those which are not, can be defined in philosophical or sociological terms. Within the latter perspective a failed discovery is simply one which goes unrecognised by the scientific community at the time it is announced. The earlier mentioned claim of having discovered element 102 belongs to the category (Sect. 3.2). In this chapter we present details of two other cases in the history of superheavy elements (SHEs), the one primarily concerning elements 112 and 122 and the other element 118.

H. Kragh, *From Transuranic to Superheavy Elements*, SpringerBriefs in History of Science and Technology, https://doi.org/10.1007/978-3-319-75813-8_4

4.1 Marinov's Superheavy Elements

In 1971 the 41-year-old Israeli physicist Amnon Marinov, together with five British physicists, reported their analysis of experiments done in CERN with 24-Gev protons hitting a target of tungsten. They suggested that they might have discovered a new SHE. Marinov was at the time known as a physicist with a solid if not impressive publication record of 26 papers in standard areas of nuclear physics published in leading journals such as *Physical Review* and *Nuclear Physics*. None of his papers until 1971 dealt with SHEs but since the 1980s he almost exclusively published on this area of physics, including laboratory-produced heavy elements as well as SHEs occurring in the natural environment. For Marinov's publications on SHEs, see http://www.marinov-she-research.com/Long-Lived-Isomeric-States.html.

The basis of the 1971 paper, the proton irradiation process, was elaborate and time-consuming as it took place over a period of nearly a year. In their papers published in *Nature* the Marinov group presented evidence for the production of element 112 by secondary reactions in the tungsten target. "We believe," they concluded, "that we may have observed the production of element 112" (Marinov et al. 1971a, b). Evidence for the possible existence of the assumedly extinct element based on analysis of anomalous xenon isotopes in meteorites had been reported two years earlier by Edward Anders, a mineralogist and meteorite specialist at the University of Chicago (Anders and Heymann 1969). He hypothesised that the measured xenon isotopes Xe-136 and Xe-137 were due to spontaneous fission of extinct SHEs in primitive meteorites. In later work Anders suggested that the meteoritic xenon anomaly might be explained by primordial elements of $Z = 113, 114$ or 115 and not by $Z = 112$ (Fontani et al. 2015, p. 371). However, the Marinov paper was the first time that element 112 was claimed to have been artificially produced.

The evidence presented by Marinov and his group was primarily in the form of observation and mass determination of spontaneous fission fragments in a mercury source chemically separated from the irradiated tungsten target. The whole argument rested on the prediction that element 112 was chemically homologous to mercury. What happened was unclear, but in subsequent papers Marinov hypothesised that some of the incident protons would knock out of the tungsten nuclei a heavy and neutron-rich nucleus at high energy (Marinov et al. 1984). This nucleus would then fuse with one of the tungsten target nuclei in a secondary reaction. He thus arrived at a "consistent interpretation" of the mass spectra in terms of reactions leading to nuclei X with $N \cong 160$, such as

$$^{86}_{38}\text{Sr} + ^{186}_{74}\text{W} \rightarrow ^{272}_{112}\text{X}$$

Marinov's group estimated the number of produced nuclei of mass number 112 to be around 500 and that they decayed by spontaneous fission with a half-life of a few weeks. Although the conclusion in the 1971 paper was cautious, Marinov had no doubt that he and his group had discovered a new SHE. In Marinov et al. (2004) he spoke of "our claim for discovering element 112" and latest in 2007 he had formally submitted a discovery claim to the Joint Working Party (JWP) established by IUPAC and IUPAP.[1] On the other hand, neither in 1971 nor in later publications did he or members of his group propose a name for the element.

Shortly after the first publication of Marinov's group a team of French physicists reported results obtained by bombarding Th-232 with Kr-84 ions. Among the reaction products the team identified alpha particles at very high energy which they suggested must be attributed to an element with $Z > 102$. "These results," one of the French physicists wrote, "can be taken as an additional indication for the existence of superheavy elements first observed by Marinov et al. in a very different experiment" (Bimbot 1971). However, this was the only experiment in apparent agreement with the claims by the Marinov group.

Most physicists and chemists engaged in SHE research were unconvinced that element 112 had been discovered. And they were equally unconvinced when Marinov, at a symposium on SHEs in Lubbock, Texas, claimed to have found evidence of $Z = 104, 111, 113$ and 114 in the form of long-lived nuclides decaying by spontaneous fission in the irradiated tungsten target (Marinov et al. 1978a,b). Some of Marinov's British collaborators reported results from more extensive experiments with a proton-irradiated tungsten target, concluding that "this experiment shows that we are unable to repeat the results in the [original] experiment" (Batty et al. 1973, p. 430). Scientists from Washington University in St. Louis conducting similar but not identical experiments also got results that disagreed with the original data reported by Marinov's group (Geisler et al. 1973). The same was the case with control experiments carried out by Flerov's group in Russia using the Serpukhov accelerator (Zhdanov 1974) and by a Swiss-German-French team using CERN's proton synchrotron (Westgaard et al. 1972). Clearly, the claim for element 112 rested on a shaky foundation, although Marinov thought otherwise.

In 1992 the Transfermium Working Group (TWG) assessed the discovery claim, tersely concluding that although "it can not be definitively dismissed," then "data reported so far are insufficient to indicate that a new element has been produced" (Barber et al. 1992, p. 507). Later assessments by the JWP did not change the verdict (Karol et al. 2001). Not only had independent attempts to duplicate the results failed, but the JWP also argued that the Marinov collaboration's interpretation in favour of $Z = 112$ rested on several assumptions of a speculative nature. The report of 2003 was unequivocally critical (Karol et al. 2003, p. 1606):

> The collaboration's arguable use of forceful expressions such as "overwhelming evidence," "clear and proven," and "impossible to refute" is neither convincing nor swaying. Extraordinary intriguing phenomena, not much selective in their measured character, are, in part, necessary for the acceptance of the collaboration's interpretations of their data.

---

[1]Minute of meeting in Torino, August 2007, of the Inorganic Chemistry Division, IUPAC, see http://old.iupac.org/divisions/II/II_Torino07min.pdf.

The JWP needs much more to be able to relinquish its deeply felt unease that the tautological rationalization of the Marinov et al. measurements remains inadequate.

The JWP repeated its critique in Barber et al. (2009).

Marinov et al. (2004) did not accept the "harsh verdict" of JWP, which the group found was inconsistent with the earlier TWG assessment. As documentation Marinov did not refer to the TWG report but, unusually, to a private letter from one of the TWG members, the French nuclear chemist Marc Lefort, from which he quoted. His main objection against the negative conclusion of the JWP was based on the claimed discovery in the late 1990s of a new form of long-lived isomeric states of atomic nuclei which he pictured as super-deformed and far from the ground state (Marinov et al. 2001). Apart from explaining the production of element 112 Marinov thought that the hypothesis of long-lived isomeric states also promised an understanding of several hitherto unexplained phenomena of nature. One of these phenomena was the controversial polonium halos claimed by Robert Gentry in the 1970s (Sect. 2.3).

As Marinov saw it, the theoretical discovery of a new type of nuclear isomers justified the interpretation in terms of $Z = 112$ and he consequently maintained the discovery claim. The German physicist Brandt (2005) supported the claim or at least considered it a possibility to be taken seriously, but his support was not considered to be of much weight as he was a member of the Marinov collaboration. As the first JWP report made clear, the hypothesis of new nuclear isomers was not an acceptable justification (Karol et al. 2001, p. 964):

[Marinov's papers] continue to press arguments for the existence of very long-lived isomeric states of actinides and transactinides and of very high fusion cross-sections for their formation, each several orders of magnitude beyond current understanding. These extraordinary phenomena are, in part, necessary for the acceptance of the collaborations' interpretations. The JWP remained unmoved.

When the special JWP report on element 112 appeared in 2009, there was no change in the verdict. Marinov and two of his collaborators responded in a lengthy paper in which they repeated and reinforced earlier arguments (Marinov et al. 2009). The paper was submitted for publication to *Pure and Applied Chemistry*, but apparently it was rejected. Contrary to what was the JWP view, so Marinov and his collaborators objected, the positive evidence for element 112 had never been refuted, for later experiments with conflicting results were not exact repetitions of the original 1971 experiment. The Marinov group and the JWP seem to have had different conceptions of the notion of repeatability.

Element 112 did become a reality, but in the version proposed by GSI (Gesellschaft für Schwerionenforschung) in Darmstadt, one of the centres of SHE research. Founded in 1969 by the state of Hessen and the West German federal government, GSI made its first experiments five years later with a linear accelerator. The establishment and early history of GSI is described in detail in Buchhaupt (1995). In 2008 the institution was renamed the GSI Helmholtz Center for Heavy Ion Research. In early 1996 a GSI team led by Sigurd Hofmann reported that two atoms of $Z = 112$ had been found by bombarding Pb-208 with Zn-70 ions

(Hofmann et al. 1996). Interestingly, the GSI authors did not refer to Marinov's earlier work. The reaction found in Darmstadt was interpreted to be

$$^{208}_{82}\text{Pb} + ^{70}_{30}\text{Zn} \rightarrow ^{277}_{112}\text{Cn} + ^{1}_{0}n$$

Although Hofmann's group claimed to have produced and identified element 112 "unambiguously," the JWP concluded in its reports from 2001 and 2003 that the evidence for the new element was insufficient. Apart from Marinov and the GSI collaboration, also Sergey Dimitriev on behalf of JINR, the Joint Institute for Nuclear Research in Dubna, claimed to have discovered the element. But in its final report the JWP judged that the Dubna experiments merely confirmed the earlier experiments in Darmstadt and thus reinforced the conclusion in favour of the latter's claim, so as did experiments conducted in Japan by Kōsuke Morita and collaborators: "The JWP has agreed that the priority of the Hofmann et al. ... collaborations' discovery of the element with atomic number 112 at GSI is acknowledged" (Barber et al. 2009, p. 1339; Bieber 2009). In agreement with the proposal of the discoverers the element was named copernicium, which was ratified by the IUPAC Council the following year.

The proposed chemical symbol was initially Cp, but it was soon pointed out that the symbol had previously been applied for "cassiopeium" or what presently is lutetium ($Z = 71$).[2] In addition, Cp was an abbreviation approved by IUPAC for the cyclopentadienyl ion in inorganic chemistry. Consequently the symbol was changed to Cn (Meija 2009). More generally IUPAC's Inorganic Chemistry Division was concerned that discoverers spread names and symbols for new elements on the internet and elsewhere before IUPAC had an opportunity to discuss them and make a decision. In this way control was slipping out of the hands of IUPAC, something the organization resisted:

> In the future, IUPAC should firmly inform any potential discoverer that any mention of names or symbols should be withheld until IUPAC grants approval. In the past, the policy directed to the Division was that this information is confidential until IUPAC makes a decision. Now this information was spread across the world, before our Division was even informed about the choice that would have to be discussed. The division views this situation as unacceptable. One consequence is that the information about what may turn out to be an unacceptable choice will exist across the scientific community for months before the correct information is agreed upon.[3]

The search for SHEs is almost synonymous with attempts to synthesise very heavy atomic nuclei, but there is the slight possibility that trace amounts of SHEs in unusual isomeric states may also occur in nature (see Sect. 2.3). This is what

---

[2]Cassiopeium, discovered by Auer von Welsbach in 1907, was for a period accepted by the German Atomic Weight Commission (but not by the International Commission) and it entered Bohr's periodic table in the early 1920s. Only in 1949 did IUPAC abandon Cp as an accepted symbol for element 71. For the complex story of the element, see Kragh (1996).

[3]Inorganic Chemistry Division committee of IUPAC, minutes of meeting at Glasgow, 2009. https://www.iupac.org/fileadmin/user_upload/divisions/II/II_Glasgow09min.pdf.

Marinov, then an emeritus professor at the Hebrew University of Jerusalem, hypothesized in papers of 2003 and 2007. The following year he and his team created attention by suggesting that they had detected—not synthesised—a long-lived element with atomic number as high as 122 in purified thorium. The paper with the discovery claim first appeared as a non-peer-reviewed *Arxiv* preprint in 2008. Marinov apparently had trouble finding a journal that would accept the paper. He submitted the paper to *Nature* which however turned it down without bothering sending it for peer review (Van Noorden 1934). In *International Journal of Modern Physics E*, a journal established in 1992, Marinov et al. (2010, p. 139) summarised:

> Mass spectral evidence has been obtained for the possible existence of a long-lived superheavy isotope with atomic mass number 292 and $t_{1/2} \geq 10^8$ y. Based on predicted chemical properties of element 122, it is probable that the isotope is $^{292}122$, but a somewhat higher Z cannot absolutely be excluded. Because of its long lifetime, it is deduced that the observed isotope at $A = 292$ is in a long-lived isomeric state rather than in the g.s. [ground state].

The abundance of the new element was estimated to be about 10^{-12} relative to Th-232. Marinov et al. (2011a) made a similar claim of having detected small amounts of a long-lived state of $Z = 111$ (roentgenium) at an abundance level ca. 10^{-10} inside a sample of gold. The half-life of natural Rg-261 was estimated to be the same as that for the nuclide of $Z = 122$, namely $t_{1/2} \geq 10^8$ y.

The possibility of synthesising element 122 in nuclear reactions had been investigated as early as 1971 in Dubna. Alexander Demin and a group of physicists bombarded targets of U-238 and Am-243 with a beam of Zn-66 ions in the hope that a fusion product would turn up. However, nothing came out of the experiment (Zhdanov 1974).

The claim of having found a monster nucleus in nature with $Z = 122$ and $N = 170$ did not meet with approval in the SHE community. Marinov had at the time a somewhat doubtful reputation and his claim was scarcely taken seriously. Rolf-Dietmar Herzberg, a nuclear physicist, judged that "the popular vote is one of ridicule" (Van Noorden et al. 2008). It cannot have added to the reputation of Marinov's work that some of his co-authors were basically amateurs with no relevant scientific record. One of them was Robert Gentry who many years earlier had claimed to have discovered element 112 in natural sources and whom Marinov had met at the 1978 Lubbock symposium on SHEs. However, by the early years of the new millennium Gentry was primarily known, if known at all, as a creationist pseudo-scientist (Sect. 2.3). Marinov took Gentry's claim of having discovered polonium halos seriously, which most geophysicists and nuclear scientists did not. Without in any way sharing Gentry's creationist ideas, in some of his papers he referred to Gentry's book *Creation's Tiny Mystery*, a creationist tract of 1986 (Marinov et al. 2001; Marinov et al. 2009).

Not only was the long half-life of 100 million years or more highly unusual and hard to explain theoretically, the experimental technique used for the identification was also criticized (Brumfiel 2008; Dean 2008). In their claim of having detected long-lived isomeric states of nuclei in thorium isotopes the Marinov group made use of a method known under the awkward abbreviation ICP-SFMS (Inductively Coupled Plasma Sector Field Mass Spectrometry) and not the better tested and more precise AMS (Atomic Mass Spectrometry) method. Using the latter method critics failed to observe the isomers. They objected that Marinov and his group had based their conclusions, including the claim of element 122, on an inappropriate instrument method. This method, they argued, did not allow the accuracy claimed by Marinov. As two of the critics summarized, the claim of having discovered long-lived isomers was "not believable" (Barber and De Laeter 2009). The Marinov collaboration rejected the arguments in Marinov et al. (2009).

The JWP seems to have ignored the latest discovery claim of the Israeli maverick physicist, except that the group may have referred to it indirectly in a report of 2016: "As serious claims associated with elements having $Z = 119$ or above have not yet been made, we note that, for the first time, the Periodic Table exists with all elements named and no proposed or pending new additions" (Öhrström and Reedijk 2016). In any case, still in his last paper, submitted to *Arxiv* in October 2011, Marinov defended the discovery of $Z = 122$ and other SHEs occurring in nature. Very precise experiments done with accelerator mass spectrometry by an Austrian research group failed to confirm the findings reported by Marinov regarding $Z = 111$ and $Z = 122$. According to the Austrian group they were probably due to "unknown artifacts" and an "unidentified background" (Dellinger et al. 2011a, b). In other words, they were wrong. Expectedly, Marinov disagreed, arguing that the apparently contradictory results did not amount to a refutation of very heavy SHEs in nature (Marinov et al. 2011b). On the contrary, he argued that the results, if properly interpreted, suggested that the Austrian group had unknowingly detected tiny amounts of Rg-296 and also of the $A = 294$ nuclide of element 115. Instead of refuting the claim of long-lived isomeric states of superheavy nuclei, the group had confirmed it.

Marinov passed away on 7 December 2011. In a later paper a German and an Austrian physicist recalled their interaction with their Israeli colleague: "Although we did not agree on his results, he was always ready to discuss our AMS experiments, looking for ways why we may have missed something which he was so convinced that it does indeed exist" (Korschinek and Kutschera 2015, p. 194). But not all have dismissed Marinov's discovery claims. There have been posthumous attempts made by his family and some of his former collaborators to rehabilitate the Israeli physicist and his claims of element discoveries. According to Dietmar Kolb, one of those collaborators, "Marinov's data have fundamentally changed our understanding of nuclei and in effect offer a new nuclear physics."[4] Recently an Israeli philosopher have sought to re-evaluate Marinov's contributions to SHE

---

[4]http://www.marinov-she-research.com/Super-Heavy-Elements-Research.html. In an interesting film placed on the internet called "Element 112, the Marinov Affair," Marinov is presented as a modern version of Galileo or Bruno, a bold and visionary scientist who became the victim of a prejudicial community of established nuclear physicists. See http://www.marinov-she-research.com/A-Documentary-Film.html.

physics, presenting him as a genius whose work had been unjustly dismissed by a "dogmatic and narrow-minded" scientific community. According to Gilead (2016), Marinov's "amazing discoveries" of new elements might well, in a fairer world, have earned him two Nobel prizes. But the rearguard attempts to change Marinov's reputation seems to have been ignored by physicists and chemists engaged in SHE research. He has been largely removed from the history of science.

4.2   The Ninov Affair

Marinov's discovery claims of SHEs were most likely unfounded and based on insufficient data and questionable interpretations. Although some of his work included features that may be characterised as "pathological" in the sense originally defined by Irving Langmuir as "the science of things that aren't so," it was not fraudulent or involving scientific misconduct (Bauer 2002). In a field as competitive and potentially rewarding as SHE one should not be surprised that the latter category turns up, as it did in connection with an early discovery claim of the element with atomic number 118.

In early June 1999 a group of fifteen researchers at LBNL (Lawrence Berkeley National Laboratory), headed by Kenneth Gregorich and including the 84-year-old SHE veteran Albert Ghiorso, announced that it had detected two new elements, one with $Z = 118$ and the other its alpha decay product with $Z = 116$ (Ninov et al. 1999). Their paper in *Physical Review Letters* was received 27 May and published in the issue of 9 August. Bombarding a lead target with a beam of Kr-86 ions accelerated to 459 meV, the group interpreted the results as due to fusion followed by an alpha decay:

$$^{208}_{82}\text{Pb} + ^{86}_{36}\text{Kr} \rightarrow ^{293}_{118}\text{X} + ^{1}_{0}n \quad \text{and} \quad ^{293}_{118}\text{X} \rightarrow ^{289}_{116}\text{Z} + ^{4}_{2}\text{He}$$

According to conventional wisdom the yield from this kind of fusion reaction would be extremely small and yet it was what the Berkeley group decided upon. The rationale behind the experiment was theoretical, namely recent calculations made by Robert Smolańczuk, a Polish physicist who at the time was a visiting scholar at LBNL. Smolańczuk's calculations were tightly connected to experiments in so far that his aim was to "propose a discovery experiment for elements 113–119 … [which] would extend considerably the Mendeleyev table" (Smolańczuk 1999). Not only did he predict that the reaction between Pb-208 and Kr-86 would occur with a relatively high cross-section, he also predicted the chain of alpha decays of the hypothetical nuclide $(A, Z) = (293, 118)$. The half-life of the first decay leading to element 116 was predicted to be between 31 and 310 ms.

The lead author of the paper in *Physical Review Letters* was 39-year-old Viktor Ninov, a Bulgarian who as a young man had fled his country and settled in West Germany to study physics at the Technical University in Darmstadt. Following his Ph.D. degree he began working in the GSI heavy-ion research group, where his skills were much appreciated. As part of the GSI team which in 1995–1996 announced the discovery of elements 110, 111 and 112, Ninov was a co-discoverer of three new elements. In 1996 he was hired by LBNL as an instrument builder and to take care of the computer program analysing the laboratory's experimental data.

Ninov, Gregorich, Ghiorso, Darleane Hoffman and their collaborators concluded from the raw data of the 1999 experiment that three instances of the reactions referred to above had been observed in the form of a sequence of alpha particles from $Z = 118$ down to $Z = 106$. Without using the term "discovery" they phrased their discovery claim as follows: "We have observed three decay chains ... consistent with the formation of $^{293}118$ and its decay by sequential α-particle emission to $^{289}116$, $^{285}114$, $^{281}112$, $^{277}110$, ^{273}Hs ($Z = 108$) and ^{269}Sg ($Z = 106$)."

The Berkeley scientists were at first sceptical with regard to Smolańczuk's theory, but it nonetheless acted as inspiration and motivation for the experiment. When they saw how well the experimental data agreed with the predictions of the Polish physicist they were pleasantly surprised. For example, the measured value of the half-life of the first alpha decay was approximately 120 µs, just in the predicted interval. "The energies of the observed α particles and their lifetimes agree remarkably well with the prediction of Smolańczuk," they wrote. What looked like a verification of a theoretical prediction added to their conviction that the data were correct and that they justified their interpretation. Ghiorso recalled that the concordance of theory and experiment made Ninov exclaiming that "Robert must talk to God" (Monastersky 2002). However, in a telephone interview of December 2006, Ninov denied that he had ever used the phrase or something like it (LeVay 2009, p. 249). Anyway, the concordance was more prosaic, as it was the result of a wrong theory and wrong data. In fact, the concordance was far from perfect as the experimental value for the Kr-Pb fusion cross-section, estimated to be approximately 2 pb (picobarn = 10^{-40} m^2), was much smaller than the ca. 670 pb calculated by Smolańczuk. The discrepancy was not highlighted in the 1999 paper.

The news from Berkeley were exciting and at first generally accepted. After all, LBNL was one of the world's leading SHE laboratories and Ninov recognised to be a talented and innovative specialist in the field. In a comment dated 13 June 1999, Ghiorso and Hoffman described enthusiastically the discovery as a "fantastic climax" in the history of SHE research: "At long last, more than 30 years after predictions of the Island of Superheavy Elements in the mid-1960s, we have indeed, reached and even gone beyond the 'magic' region around $Z = 114$ and $N = 184$! ... Now there is no question, the SuperHeavy Island actually exists" (Hoffman et al. 2000, pp. 426–431). News of the discovery was spread by *New York Times* (Browne 1999), announcing that "In a discovery that came as a complete surprise to most nuclear chemists, an international team of scientists at Lawrence Berkeley National Laboratory in California has added two new elements to the periodic table." Unfortunately, what a first appeared to be a great success soon turned into something like a nightmare.

Neither the Darmstadt group nor laboratories in France (GANIL Heavy-Ion Research Laboratory) and Japan (RIKEN) were able to confirm the data on the reported synthesis of the nuclide of mass number 293. And when the Berkeley scientists tried to repeat and improve the experiments in early 2000, no sign of element 118 turned up. Still late in March, when presenting her Priestley Medal Address at a ceremony of the American Chemical Society, Darleane Hoffman referred confidently to the new elements 118 and 116, which "suddenly and rather unexpectedly ... have burst upon the scene" (Hoffman 2000). But later in the year it dawned upon the Berkeley group that something might be wrong and perhaps seriously wrong.

In June 2001 a review committee headed by Hoffman was set up to examine the matter and a few months later it was replaced by a formal investigation committee chaired by Rochus Vogt, a retired Caltech physicist. In a detailed internal report of March 2002 the Vogt committee stated that "There is clear evidence to conclude that Dr. Ninov has engaged in misconduct in scientific research by carrying out this fabrication. ... If anyone else had done the fabrication, Dr. Ninov would almost surely have detected it." The Vogt committee further found it "incredible that no one in the group, other than Ninov, examined the original data to confirm the purported discovery of element 118" (Goodstein 2010, p. 105; Schwarzschild 2002). Although Ninov vigorously denied the charges, in November 2001 he was put on indefinite paid leave and in May 2002 fired from LBNL. He subsequently got a teaching position at the University of the Pacific in Stockton, California, but after 2006 he seems to have vanished from the academic world (LeVay 2009, p. 250).

In emails of 27 July 2001 to his colleagues in the SHE community, Gregorich apologised as leader of the Berkeley group for the erroneous report and the undesirable consequences it might have caused (Hofmann 2002, p. 201). At the same time he sent a letter of retraction to *Physical Review Letters*, but the editors declined publishing it because Ninov, who insisted to have his name removed, was not among the authors. It only appeared a year later. "We conclude that the three reported chains are not in the 1999 data [and] retract our published claim for the synthesis of element 118," announced the fourteen remaining authors (*Physical Review Letters* **89**: 039901). The 6-line retraction gave the impression of a simple mistake, not a case of fraud, and it did not mention Ninov by name. Maintaining his innocence Ninov felt no need to apologise. On 5 September 2002 he said on the National Public Radio: "So far I wasn't able to discover a mistake from my side and I disagree with the laboratory that I fabricated the data because I just simply didn't have a motivation for this. I mean, what was my profit or would have been my profit?" (*APS News* **11**, October 2002). And in a conversation from the same time reported in Monastersky (2002): "I'm the scapegoat. I don't accept the accusation of fabricating the data. They're saying I'm smart enough to run a complicated experiment, but I'm dumb enough not to fake it properly."

What was immediately considered an embarrassing case of scientific misconduct attracted scientific as well as public interest. Contemporary accounts include Hofmann (2002, pp. 194–204), Johnson (2002), Schwarzschild (2002), Monastersky (2002), *Nature* **420** (2002): 728, *Science* **293** (2001): 777 and **297** (2002): 313, and *New York Times* (23 July 2002). The CNIC report of 2003 briefly referred to the Berkeley claim and its retraction (Karol et al. 2003).

The Ninov case was exposed at about the same time as the "Schön scandal," a reference to the German physicist Jan Hendrik Schön who was fired from the Bell Labs for having manipulated and fabricated data in semiconductor research (Goodstein 2010, pp. 97–102). The two cases led to much soul-searching and did much to raise awareness of scientific misconduct in the physical sciences (Kirby and Houle 2004). As pointed out in *Chemical and Engineering News*, a weekly ACS journal, fraud of this kind should not happen, and yet it did—twice (Jacoby 2002).

At a LBNL meeting for employees on 25 June 2001, the laboratory's director Charles Shank said that "all coauthors have a responsibility before a paper is published to verify the data. In this case, the most elementary checks and data achieving were not done" (Muir 2002). As mentioned, the Vogt committee repeated the allegation nine months later. However, the suggestion of a common responsibility was resisted by Gregorich and some of the other involved physicists who felt that it was unjustified. It raised the question of who were responsible for what in collaborative multi-author experimental work. Could it be reasonably claimed that all co-authors carry full responsibility for the content of a scientific paper? As pointed out by Trilling (2003), a Berkeley high-energy physicist and member of the investigation committee, this was unrealistic since "many experimental papers … go unread by a significant fraction of the co-authors, who can often total several hundred."[5] As the best and perhaps only guarantee against erroneous results, Trilling referred to confirmation by independent research teams. This was also the solution that Shank fell back on. "Science is self-correcting," he claimed. "If you get the facts wrong, your experiment is not reproducible. In this case, not only did subsequent experiments fail to reproduce the data, but also a much more thorough analysis of the 1999 data failed to confirm the events" (Anon. 2001).

Disregarding the philosophical question of whether science is self-correcting or not, it is of interest to recall that the TWG in its 1991 report on criteria of discovery had addressed the issue of repeatability in a somewhat different way. Although the TWG stressed the importance of independent confirmation, it also realised that it was not a magic wand and that in some cases the repeatability or confirmation criterion could be dispensed with (Wapstra 1991, p. 883). There are several cases in the history of science of experimental confirmation of wrong theories, which may happen for a variety of causes. In this case it is unclear if Ninov was biased toward Smolańczuk's theory and wanted to confirm it, but most likely the reasons behind his data fabrication were different. Brush (2015, p. 17) suggests that Ninov may unconsciously have wanted to confirm Smolańczuk's prediction, but this is nothing but a speculation.

Not only had the claims of elements 116 and 118 evaporated, scientists at GSI were also forced to reconsider their own claim of having discovered elements 111

---

[5]In experimental high-energy physics the number of authors may count several thousand. A paper on the mass of the Higgs boson appearing in *Physical Review Letters* **114** (2015): 191803, lists no less than 5,154 authors! It is hard to imagine that all authors read the paper critically and attentively if at all.

and 112. Evidence for these elements had been found in experiments from the mid-1990s in which Ninov, then at GSI, was responsible for the data analysis. According to a 2002 paper by Hofmann and collaborators, "In two cases ... we found inconsistency of the data, which led to the conclusion, that for reasons not yet known to us, part of the data used for establishing these two [decay] chains were spuriously created" (Hofmann et al. 2002, p. 156). Ten years later, looking back at the development, two GSI veterans still wrote as if the name of the forger was a mystery. But they also wrote that "The faking of results started at GSI and was exported to LBNL" (Armbruster and Münzenberg 2012, p. 298). Fortunately, the basis for the German discovery claims of the two elements was not affected by the spurious data. The GSI priority to elements 111 and 112 was recognised by CNIC in 2003 and 2009 respectively. Element 116 (livermorium) won recognition in 2011 when priority to the discovery was assigned to a collaboration of Dubna and Livermore scientists (Sect. 6.1).

And what about $Z = 118$? For how long would it remain ununoctium or eka-radon? In early experiments made in Dubna between 2002 and 2004 the scientists identified two events from bombarding Cm-245 with Ca-48 that could be ascribed to the decay of a nuclide of element 118, but the evidence was still not convincing (Oganessian et al. 2004). Further experiments of 2006 led to greater confidence (Oganessian et al. 2006). The new results were reported by a Russian-American collaboration headed by the Dubna physicist Yuri Oganessian, a highly esteemed veteran in SHE research. The big-science team consisted of 30 members, of whom 20 were from JINR and 10 from LLNL (Lawrence Livermore National Laboratory). The team observed three decay chains arising from the fusion of Ca-48 and Cf-249, which they interpreted as

$$\ce{^{48}_{20}Ca + ^{249}_{98}Cf -> ^{294}_{98}Z + 3^{1}_{0}n}$$

The formed nuclei were primarily identified by alpha decays with a half-life of only 0.9 ms. Oganessian and his co-authors briefly mentioned the 1999 LBNL experiment (which "was later disproved"), but without referring to Ninov or the retracted paper. In its 2011 report JWP concluded that the evidence obtained by the JINR-LLNL collaboration did not yet satisfy the criteria for discovery. Consequently the collaboration undertook to confirm and improve the results in subsequent experiments, but it took until 2016 before the JPW concluded that the work reported ten years earlier satisfied the criteria for discovery (Karol et al. 2016). The provisional name ununoctium was now replaced with oganesson in recognition of the leader of the discovery team. The chemical symbol of the element, to this day the heaviest of all, became Og. The ending "-on" in the name was in accordance with the IUPAC recommendations for elements in group 18, such as xenon and radon (Koppenol et al. 2016). Likewise, elements in group 17 shall end on "-ine" as in chlorine and iodine, and consequently element 117 was named tennessine.

References

Anders, E., Heymann, D.: Elements 112 to 119: were they present in metorites? Science **164**, 821–823 (1969)

Anon.: Results of element 118 experiment retracted. Berkeley Lab Research News, 27 July 2001. http://enews.lbl.gov/Science-Articles/Archive/118-retraction.html

Armbruster, P., Münzenberg, G.: An experimental paradigm opening up the world of superheavy elements. Eur. Phys. J. H **37**, 237–309 (2012)

Barber, R.C., et al.: Discovery of the transfermium elements. Prog. Part. Nucl. Phys. **29**, 453–530 (1992)

Barber, R.C., et al.: Discovery of the element with atomic number 112. Pure Appl. Chem. **81**, 1331–1343 (2009)

Barber, R.C., De Laeter, J.R.: Comment on 'Existence of long-lived isomeric states in naturally-occurring neutron-deficient Th isotopes'. Phys. Rev. C **79**, 049801 (2009)

Batty, C.J.: Search for superheavy elements produced by secondary reactions in a tungsten target. Nature **244**, 429–430 (1973)

Bauer, H.H.: 'Pathological science' is not scientific misconduct (nor is it pathological). Hyle **8**, 5–20 (2002)

Bieber, C. (2009): Element 112 joins the periodic table. New Sci. **202**, 10, 20 June 2009

Bimbot, R.: Complete fusion induced by krypton ions: indications for synthesis of superheavy nuclei. Nature **234**, 215–216 (1971)

Brandt, R.: Comments on the question of the discovery of element 112 as early as 1971. Kerntechnik **70**, 170–172 (2005)

Browne, M.W.: Team adds 2 new elements to the periodic table. New York Times, 9 June 1999

Brumfiel, G.: The heaviest element yet? Nature, 1 May 2008. http://www.nature.com/news/2008/080501/full/news.2008.794.html

Brush, S.G.: Making 20th Century Science: How Theories Became Knowledge. Oxford University Press, Oxford (2015)

Buchhaupt, S.: Die Gesellschaft für Schwerionenforschung. Campus Verlag, Frankfurt am Main (1995)

Dean, T.: Higher, higher! New Sci. **199**(2666), 32–35 (2008)

Dellinger, F., et al.: Ultrasensitive search for long-lived superheavy nuclides in the mass-range $A = 288$ to $A = 300$ in natural Pt, Pb, and Bi. Phys. Rev. C **83**, 065806 (2011a)

Dellinger, F., et al.: Upper limits for the existence of long-lived isotopes of roentgenium in natural gold. Phys. Rev. C **83**, 015801 (2011b)

Fontani, M., Costa, M., Orna, M.V.: The Lost Elements: The Periodic Table's Shadow Side. Oxford University Press, Oxford (2015)

Geisler, F.H., Philips, P.R., Walker, R.M.: Search for superheavy elements in natural and proton-irradiated materials. Nature **244**, 428–429 (1973)

Gilead, A.: Eka-elements as chemical pure possibilities. Found. Chem. **18**, 183–194 (2016)

Goodstein, D.: On Fact and Fraud: Cautionary Tales from the Front Lines of Science. Princeton University Press, Princeton (2010)

Hoffman, D.C.: The new millennium. Chem. Eng. News **78**(13), 36–42 (2000)

Hoffman, D.C., Ghiorso, A., Seaborg, G.T.: Transuranium People: The Inside Story. Imperial College Press, London (2000)

Hofmann, S.: On Beyond Uranium: Journey to the End of the Periodic Table. Taylor & Francis, London (2002)

Hofmann, S., et al.: The new element 112. Z. Phys. A **354**, 229–230 (1996)

Hofmann, S., et al.: New results on elements 111 and 112. Eur. Phys. J. A **14**, 147–157 (2002)

Jacoby, M.: Fraud in the physical sciences. Chem. Eng. News **80**, 31–33 (2002)

Johnson, G.: At Lawrence Berkeley, physicists say a colleague took them for a ride. New York Times D1, 15 October 2002

Karol, P.J., et al.: On the discovery of the elements 110–112. Pure Appl. Chem. **73**, 959–967 (2001)

Karol, P.J., et al.: On the claims for discovery of elements 110, 111, 112, 114, 116, and 118. Pure Appl. Chem. **75**, 1601–1611 (2003)

Karol, P.J., et al.: Discovery of the element with atomic number $Z = 118$ and completing the 7th row of the periodic table. Pure Appl. Chem. **88**, 155–160 (2016)

Kirby, K., Houle, F.A.: Ethics and the welfare of the physics profession. Phys. Today **57** (November), 42–46 (2004)

Koppenol, W.H., et al.: How to name new elements. Pure Appl. Chem. **88**, 401–405 (2016)

Korschinek, G., Kutschera, W.: Mass spectrometric searches for superheavy elements in terrestrial matter. Nucl. Phys. A **944**, 190–203 (2015)

Kragh, H.: Elements no. 70, 71 and 72: Discoveries and controversies. In: Evans, C.H. (ed.) Episodes from the History of the Rare Earth Elements, pp. 67–90. Kluwer Academic, Dordrecht (1996)

LeVay, S.: When Science Goes Wrong: Twelve Tales From the Dark Side of Discovery. Monday Books, Reading (2009)

Marinov, A., et al.: Evidence for the possible existence of a superheavy element with atomic number 112. Nature **229**, 464–467 (1971a)

Marinov, A., et al.: Spontaneous fission previously observed in a mercury source. Nature **234**, 212–215 (1971b)

Marinov, A., Eshar, S., Weil, J.L.: Production of actinides by secondary reactions in the bombardment of a tungsten target with 24 GeV protons. In: Lodhi, M. (ed.) Superheavy Elements. Proceedings of the International Symposium on Superheavy Elements, pp. 72–80. Pergamon, New York (1978a)

Marinov, A., Eshar, S., Aspector, B.: Study of Au, Tl and Pb sources separated from tungsten targets that were irradiated with 24 GeV protons, indications for the possible production of superheavy elements. In: Lodhi, M. (ed.) Proceedings of the International Symposium on Superheavy Elements, pp. 81–88. Pergamon, New York (1978b)

Marinov, A., Gelberg, S., Kolb, D.: Abnormal radioactive decays out of long-lived super- and hyper-deformed isomeric states. Acta Phys. Hung. **13**, 133–137 (2001)

Marinov, A., et al.: Consistent interpretation of the secondary-reaction experiments in W targets and prospects for production of superheavy elements in ordinary heavy-ion reactions. Phys. Rev. Lett. **52**, 2209–2212 (1984)

Marinov, A., et al.: Response to the IUPAC/IUPAP joint working party second report (2004). arXiv:nucl-ex/0411017

Marinov, A., et al.: Reply to 'Comment on "Existence of long-lived isomeric states in naturally-occurring neutron-deficient Th isotopes"'. Phys. Rev. C **79**, 0498802 (2009)

Marinov, A., Kolb, D., Weil, J.L.: Response to 'Discovery of the element with atomic number 112' (2009). arXiv:nucl-ex/09091057

Marinov, A., et al.: Evidence for the possible existence of a long-lived superheavy nucleus with atomic mass number $A = 292$ and atomic number $Z \cong 122$ in natural Th. Int. J. Mod. Phys. E **19**, 131–140 (2010)

Marinov, A., et al.: Enrichment of the superheavy element roentgenium (Rg) in natural Au. Int. J. Mod. Phys. E **20**, 2391–2401 (2011a)

Marinov, A., et al.: ICP-SFMS search for long-lived naturally-occurring heavy, superheavy and superactinide nuclei compared to AMS experiments. Int. J. Mod. Phys. E **20**, 2403–2406 (2011b)

Meija, J.: The need for a fresh symbol to designate copernicium. Nature **461**, 341 (2009)

Monastersky, R.: Atomic lies. The Chronicle of Higher Education, 16 August 2002. http://www.chronicle.com/article/Atomic-Lies/31825

Muir, H.: Elementary mistake due to falsified data. New Sci., 15 July 2002. https://www.newscientist.com/article/dn2545-elementary-mistake-due-to-falsified-data/

Ninov, V., et al.: Observation of superheavy nuclei produced in the reaction of Kr-86 with Pb-208. Phys. Rev. Lett. **83**, 1104–1107 (1999)

Öhrström, L., Reedijk, J.: Names and symbols of the elements with atomic numbers 113, 115, 117 and 118. Pure Appl. Chem. **88**, 1225–1229 (2016)

Oganessian, YuT, et al.: Heavy element research at Dubna. Nucl. Phys. A **734**, 109–123 (2004)

Oganessian, YuT, et al.: Synthesis of the isotopes of elements 118 and 116 in the ^{249}Cf and ^{245}Cm + ^{48}Ca fusion reactions. Phys. Rev. C **74**, 044602 (2006)

Schwarzschild, B.: Lawrence Berkeley lab concludes that evidence of element 118 was a fabrication. Phys. Today **55**(September), 15–17 (2002)

Smolańczyk, R.: Production of superheavy elements. Phys. Rev. C **60**, 21301 (1999)

Trilling, G.: Co-authors are responsible too. Phys. World **16**(June), 16 (2003)

Van Noorden, R.: Heaviest element claim criticised. Chem. World, 2 May 2008. https://www.chemistryworld.com/news/heaviest-element-claim-criticised/3001934.article

Wapstra, A.H.: Criteria that must be satisfied for the discovery of a new chemical element to be recognized. Pure Appl. Chem. **63**, 879–886 (1991)

Westgaard, L.: Search for super-heavy elements produced by secondary reactions in uranium. Nucl. Phys. A **192**, 517–523 (1972)

Zhdanov, G.B.: Search for transuranium elements (methods, results, and prospects). Sov. Phys. Usp. **16**, 642–658 (1974)

Chapter 5
The Transfermium Wars

Abstract With the rival discovery claims in the early 1970 of the first transactinide elements, numbers 104 and 105 in the periodic table, the situation with regard to official recognition of new elements became increasingly chaotic. Definite criteria for discovery and naming procedures were needed. As a result, a Transfermium Working Group (TWG) was established in 1985 jointly by IUPAC and IUPAP. The TWG reports of 1991 and 1992, including discovery criteria as well as evaluations of specific discovery claims, were controversial and central parts of what in some quarters became known as the "transfermium wars." The so-called wars were basically an extended priority dispute which lasted until the late 1990s. The name of element 106, eventually settled to be seaborgium, played an important role in the transfermium warfare. Although the Cold War had officially ended, its shadows were visible in the rivalry between American and Russian scientists.

Keywords Discovery criteria · Priority disputes · IUPAC · Element names Seaborgium · Transfermium working group

What has been called the "transfermium wars" refer to a series of convoluted disputes concerning the names and discoveries of primarily the elements with atomic numbers from 104 to 109 but also including a few other elements. The pugnacious name may first have been used by Paul Karol, who in 1994 commented critically on the ongoing naming controversy (Karol 1994; Rothstein 1995). A member of the ACS Committee on Nomenclature and later serving as Chair of the Joint Working Party, Karol was centrally involved in this and other naming and priority controversies regarding the heavy synthetic elements.

H. Kragh, *From Transuranic to Superheavy Elements*, SpringerBriefs in History of Science and Technology, https://doi.org/10.1007/978-3-319-75813-8_5

5.1   A Brief Tour from $Z = 104$ to $Z = 109$

The earliest claim of having detected element 104, eventually named rutherfordium (Rf), appeared in 1964 when researchers at JINR (Joint Institute for Nuclear Research) in Dubna reported to have produced the element when bombarding Pu-242 with Ne-22 ions (Seaborg and Loveland 1990, pp. 51–64; Hoffman et al. 2000, pp. 263–282). Their only evidence was an isotope that underwent spontaneous fission. However, the Dubna experiment did not clearly identify the atomic mass of the produced nuclide and nor did it accurately determine its half-life. For these and other reasons it was not considered a convincing identification of the new element. Five years later a team of Berkeley scientists led by Albert Ghiorso provided conclusive evidence, or what they claimed was conclusive evidence, by synthesising $Z = 104$ from californium and carbon:

$$^{12}_{6}C + ^{249}_{98}Cf \rightarrow ^{257}_{104}Rf + 4^{1}_{0}n$$

A similar synthesis with C-13 projectiles resulted in the nuclide of mass number 259. The atomic number was in both cases identified by detecting the nobelium daughter nuclides No-253 and No-255 resulting from the alpha decays. In 1973 a group of physicists at Oak Ridge National Laboratory independently confirmed the atomic number of Rf-257 by means of X-ray spectroscopy.

Also element 105 (dubnium, Db) was first reported by scientists at the JINR, who in 1968 and again in early 1970 announced the identification based on an analysis of a nuclear reaction between Am-243 and Ne-22 ions. Later in 1970 the Ghiorso team reported experiments which questioned some of the Dubna results and independently resulted in a claim for a nuclide of element 105. The reported Berkeley reaction was

$$^{15}_{7}N + ^{249}_{98}Cf \rightarrow ^{260}_{105}Db + 4^{1}_{0}n$$

At a meeting of the American Physical Society on 27 April 1970 Ghiorso confidently announced the discovery of the new element, which he proposed to call hahnium in honour of Otto Hahn, the German radiochemist, Nobel Prize laureate and co-discoverer of uranium fission. But further work by the JINR group seemed to confirm that nuclides of mass number either 260 or 261 had indeed first been found in Dubna. They suggested the clumsy name nielsbohrium, symbol Ns, for the element.

Certainly, according to Georgii Flerov and his Dubna group priority to the two elements belonged to them. In a contemporary review paper Flerov (1970a, p. 394) concluded as follows:

> The synthesis of element 104 and the observation of its decay was first established in 1964 in Dubna. ... The Nuclear Reactions Laboratory of the Joint Institute for Nuclear Research in Dubna, in which teams of scientists from the Socialist countries are carrying out

investigations on the transuranium elements, counts among its assets the discovery of the three heavy elements 102, 103, and 104.

Again, at an IUPAC conference in Hamburg in September 1973, the Dubna nuclear chemist Ivo Zvára concluded that further work on element 104 "has proved once again that spontaneous fission of ^{259}Ku [kurchatovium-259] was observed in the 1966 Dubna chemical experiments" and thus established the discovery of the isotope (Zvára 1973).

Although the Russians had not initially proposed a name for the element, Flerov (1970b) reminded his American colleagues that, "unfortunately, there are examples in the history of synthesizing new elements when haste in the announcement of a discovery and naming a new element has led to a situation where a little while after the sensation only the name was left, but the nature of it was radically revised (please recall the history of element 102)." Flerov's paper was a sharp response to the discovery claim reported in Holcomb (1970) and according to which "The Soviets have not proposed a name for the element, so they apparently do not feel that their experimental evidence is very strong." In a rejoinder to Flerov's paper, Ghiorso (1971) apologised for not having referred to the newest Dubna results but otherwise maintained that only the Berkeley experiments demonstrated beyond any doubt the existence of element 105. By the early 1970s, after further work in Berkeley and Dubna, element 105 was thought to have been discovered. But it was still unclear who had discovered it and if the discovery was conclusively established. And, not least, what were the criteria for a conclusive discovery of a new element?

The case of $Z = 106$ (seaborgium, Sg) was even more controversial with regard to both discovery and name, and again it involved an extensive dispute between Berkeley and Dubna scientists. Referring to this element and also to the discovery of elements 104 and 105, Seaborg and Loveland (1990, p. 51) wrote of a "considerable controversy, frequently punctuated by acerbic comments."

The first evidence of element 106 was reported in 1974 by a Dubna team led by Flerov and Yuri Oganessian, the main evidence being events of spontaneous fission of what was supposed to be the nuclide of mass number 259. The nuclide was reportedly produced by bombarding lead isotopes with Cr-54. Shortly thereafter a team of scientists from Berkeley and the Lawrence Livermore National Laboratory (LLNL), among them Ghiorso and Seaborg, produced the $A = 263$ nuclide by bombarding a Cf-249 target with O-18 ions. They identified the atomic number of the nuclide by the alpha decays of the new nuclide and its daughter nuclide:

$$^{263}_{106}\text{Sg} \rightarrow {}^{259}_{104}\text{Rf} + {}^{4}_{2}\text{He} \quad \text{and} \quad {}^{259}_{104}\text{Rf} \rightarrow {}^{255}_{102}\text{No} + {}^{4}_{2}\text{He}$$

Experiments made in Dubna in 1984 showed that the spontaneous fission activity originally assigned to the $A = 259$ nuclide of element 106 was incorrect and in reality due to nuclides of element 104. This meant in effect a retraction of the earlier discovery claim. The case of element 106 is covered in Hoffman et al. (2000, pp. 300–327) and it will be further considered in Sect. 5.3.

To proceed numerically, for once the Berkeley scientists were not centrally involved in the discovery of element 107 (bohrium, Bh). Once again Oganessian's Dubna group came first with a cautious discovery claim dating from 1976 and which also included the isotope $A = 257$ of element 105. But as far as element 107 was concerned, the Russians were wrong as the spontaneous fission of Bh-261 on which they based their claim, was never confirmed. The element, in the form of the alpha-active nuclide with $A = 262$, was only definitely detected in Darmstadt experiments from 1981 led by Peter Armbruster and Gottfried Münzenberg (Armbruster 1985; Armbrüster and Münzenberg 2012). These so-called "cold fusion" experiments led the group to identify four events as decays of Bh-262 arising from the reaction

$$^{209}_{83}\text{Bi} + ^{54}_{24}\text{Cr} \rightarrow ^{262}_{107}\text{Bh} + ^{1}_{0}n$$

Early heavy-ion experiments producing superheavy elements (SHEs) were "hot fusion" between actinide targets and light heavy ions in which the SHE nucleus was produced in a highly excited state. The alternative "cold fusion" method of producing nuclei at low excitation energy and with the lowest possible bombardment energy was introduced by Oganessian in the 1970s. In this kind of fusion reaction the projectiles are relatively heavy ions and the target nuclides near the doubly magic Pb-208 nuclide. The technique proved very important in the production of heavy elements beyond $Z = 106$. While cold fusion quickly became the favoured method in Dubna and Darmstadt, for some years the method was regarded with suspicion in Berkeley. It needs to be pointed out that the SHE cold fusion has only the name in common with the sensational claim in 1989, made by Martin Fleischmann and Stanley Pons, of having produced hydrogen-to-helium fusion at room temperature by electrochemical means (Huizenga 1992).

Also the next two elements were discovered in Darmstadt, although not without competition. What became known as hassium (Hs, $Z = 108$) was discovered by the GSI team in Darmstadt in 1984 in reactions using Fe-58 ions bombarding a lead target. Observation of the decay of nuclides of element 108 had a little earlier been reported by Oganessian's group in Dubna but the discovery claim was considered less convincing and tacitly withdrawn. The delicacy of the GSI experiments are illustrated by the number of produced atoms sufficient for the identification of element 108: in one experiment three atoms of Hs-265 were produced and in another experiment just a single atom of Hs-264. Finally, element 109 (meitnerium, Mt) was identified by the GSI scientists before they discovered element 108. Bombarding Bi-209 with Fe-58 ions produced a single atom of the new element:

$$^{209}_{83}\text{Bi} + ^{58}_{26}\text{Fe} \rightarrow ^{266}_{109}\text{Mt} + ^{1}_{0}n$$

The discovery event appeared in 1982 and a full paper on the new element was published six years later. The detected nuclide Mt-266 was alpha radioactive with a half-life of approximately 5 ms.

5.2   Criteria for Discovery

The confusing number of discovery claims for new SHEs through the 1960s and 1970s inevitably caused reconsideration of the old question, what does it mean to have discovered a new element? The answer to the question implied answers to priority claims and also, since it was generally agreed that the discoverers had the right to propose a name, to the names of the new elements. Reluctant to enter into priority claims or making judgements about the reliability of scientific work, IUPAC's Commission on Nomenclature of Inorganic Chemistry (CNIC) maintained that its function was limited to nomenclature. It did make proposals about the naming of the new elements, but only referred to the nomenclature problem in general terms. In a minute from a meeting in 1969 the problem was explained as follows (Fennell 1994, p. 269; see also Fernelius et al. 1971):

> It had become abundantly clear during the past year that nuclear physicists reporting the preparations of new elements were exceedingly strongly attached to the right of the discoverer to select a name, and reluctant to accept the Commission's principle that whilst early selection of a name was a matter of convenience it carried no implications regarding priority of discovery.

CNIC suggested that new elements should not be named until five years had elapsed after the first discovery claims. "The period would, it was hoped, allow confirmation of the initial discovery in another laboratory, and preferably in another country." American and Russian SHE scientists generally ignored the suggestion and seem to have had little respect for the work of CNIC and IUPAC generally. As the Dubna group pointed out, somewhat condescendingly, the members of the commission "have never been and are not now experts in these fields of science [nuclear physics and radiochemistry]" (Flerov et al. 1991, p. 453).

On the proposal of Dubna, in 1974 IUPAC in collaboration with IUPAP appointed an ad hoc group of nine supposedly neutral experts chaired by Jack Lewis, a chemistry professor at Cambridge University. Three of the members were from the United States and three from the Soviet Union. The purpose of the group was "to consider the claims of priority of discovery of elements 104 and 105 and to urge the laboratories at Berkeley (USA) and Dubna (USSR) to exchange representatives regarding these experiments."[1] However, the initiative was a failure as the committee never completed its work or issued a report. Indeed, it never met as a group (Wapstra 1991, p. 881). According to Fennell (1994, p. 269), "In 1977 IUPAP said it had lost interest as the existence of the two elements was doubtful anyway." The more definite reasons for the failure are not known, but it seems that

---

[1]Quoted in Hyde et al. (1987). The three authors were the US members of the 1974 group and their paper was originally intended to be a contribution to the working group from the perspective of the Americans. The representatives from the USSR were V. I. Goldanski, S. P. Kapitza and B. M. Kedrov, and from the neutral countries A. Baumgartner from Switzerland and U. Stille from West Germany.

communication problems with the American and Russian laboratories were in part responsible. Of course, these problems were only aggravated by the Cold War.

The pros and cons of different methods for identifying the atomic number of new SHEs were discussed in Thompson and Tsang (1972). In a paper in *Science* four years later a group of Western SHE specialists, including Glenn Seaborg from Berkeley and Günter Herrmann from Darmstadt, pointed out that lack of definite discovery criteria "has contributed significantly to the competing claims for the discovery for these [transuranic] elements" (Harvey et al. 1976). The authors included seven Americans, one German and one Frenchman but none from the Soviet Union. In discussing various ways of identifying new elements, some chemical and other physical, the authors emphasised proof of the atomic number as essential. On a more practical note, "any claim to such a discovery should be published in a refereed journal with sufficient data to enable the reader to judge whether the evidence is consistent with such criteria." And as to the name of a new element, it "should not be proposed by the discoverers until the initial discovery is confirmed." The 1976 article was general in nature and did not apply the proposed discovery criteria to the ongoing controversy. But several years later three of the American co-authors published a detailed investigation of the discovery claims of elements 104 and 105 which was largely based on the 1976 criteria (Hyde et al. 1987).

The authors of the *Science* paper distinguished between two kinds of nuclear-physical methods for the identification of a new element. The one was proof of a genetic decay relationship through a chain of alpha-decaying nuclei, and the other was proof of spontaneous fission and measurement of its lifetime. While they judged the first method, the one favoured by the Berkeley group, to be acceptable, the second method, which was widely used in Dubna, "cannot per se establish that an element with a new atomic number has been produced." Both groups recognised the experimental difficulties, but the Dubna scientists were more inclined to consider spontaneous fission directly useful for the identification of new elements. At an international symposium on new elements held in Dubna in September 1980 Oganessian and his JINR colleague Yuri Lazarev expressed some scepticism with regard to the alpha decay method and they objected to what they thought was a downgrading of the spontaneous fission method in the 1976 *Science* paper. Alluding to this paper and its "a priori elaboration of the criteria that should be met to prove the discovery of a new natural phenomenon," they labelled such criteria as "ideology" (Oganessian and Lazarev 1981, p. 941).

And yet the Dubna nuclear scientists shared with their Western colleagues the concern over the confusion caused by missing criteria for SHE discoveries. As early as 1971 Flerov and his Dubna colleague, the Czechoslovakian nuclear chemist Ivo Zvára, had written a memorandum in which they pointed out that the concept of element belonged to chemistry and atomic physics and not to nuclear physics. The memorandum was in Russian and is here quoted from the English version in Flerov et al. (1991). According to the two authors:

> If only the radioactive properties of isotopes are studied and nuclear-physics proofs of the correctness of the identification are used, the work can be viewed as the discovery of an element only when the conclusions about both the atomic number and the mass number are not revised in subsequent studies. ... If the atomic number is established by chemical means or by techniques of atomic physics (Roentgen spectroscopy, etc.), then even without a nuclear-physics identification the work should be considered a discovery. In this case the mass number of the isotope may remain completely unknown.

Flerov and Zvára further stated that discovery claims were only acceptable if they appeared in print and included sufficient experimental details. In later papers the Dubna group argued that the right to name an element should belong to the authors whose contributions were decisive for the discovery (Flerov et al. 1991). But who were these authors? The different SHE groups obviously disagreed, which Flerov and his colleagues ascribed to the absence of consistent criteria for evaluating priority claims. Such criteria were badly needed. They were, after all, more than just ideology.

Shortly before the establishment of the TWG, described in Sect. 3.1, the leading GSI physicist Peter Armbruster wrote a review article in which he suggested a rule for the naming of SHEs. At the same time he expressed his wish that the many controversies over names were taken care of by an international commission of physicists and nuclear chemists. The proposal of Armbruster (1985, p. 175), which probably inspired the not yet established TWG, was this:

> Element synthesis becomes production of a given isotope, and a name should be accepted only if the experiment claiming the discovery is reproducible. An isotope is defined by its mass and atomic number, its fingerprints are its decay modes and its half-life. ... The proposed rule should be applied retrospectively for all elements discovered by isotope identification, that is elements 102-109.

The rule was not primarily about names, but more about criteria for discovery. Its emphasis on reproducibility was reflected in the later TWG criteria, but in a less rigid formulation. Indeed, it is a problematic concept and Armbruster did not explain precisely what he had in mind when claiming that experiments must be reproducible.

In a report published in *Pure and Applied Chemistry* in 1991 the TWG investigated systematically and thoughtfully criteria for recognising the existence of a new chemical element (Wapstra 1991, p. 882). This was the first part of the TWG report, dealing with criteria that must be satisfied for a discovery claim of a new element to be recognised as a discovery. The second part, an evaluation of various discovery claims of transfermium elements and conclusions regarding priority, appeared in Wilkinson et al. (1993). Both parts were reprinted in Barber et al. (1992) published in a physics journal and not in *Pure and Applied Chemistry*, a journal which was not widely known in the nuclear physics community.

The TWG fully realised the complexity of the discovery concept, implying that it could not be codified in a simple way. As shown by the history of element discoveries there had always been dissenting views of when and by whom an element was discovered, and the group argued that this was inevitable and had to be accepted (Wapstra 1991, p. 862):

Different individuals or different groups may take different views as to the stage in the accumulation of evidence at which conviction is reached and may take different views as to the existence or otherwise of crucial steps leading to that conviction and as to which those crucial steps were. Such differences can be perfectly legitimate scientifically, in that they may depend upon, for example, differing views as to the reliability of the inference that might be drawn from certain types of evidence, while not disputing the reliability of the evidence itself. So, although the scientific community may reach consensus as to the existence of a new element, the reaching of that consensus is not necessarily a unique event and different views may, in all scientific honesty, be taken as to the steps by which it was reached.

Nonetheless, the TWG came up with the following summary definition: "Discovery of a chemical element is the experimental demonstration, beyond reasonable doubt, of the existence of a nuclide with an atomic number Z not identified before, existing for at least 10^{-14} s." With regard to the requirement of a minimum lifetime of the nuclide it was introduced to make the formula more chemical and in accord with the standard view of the term element: "It is not self-evident that 'element' makes sense if no outer electrons, bearers of the chemical properties, are present." It takes about 10^{-14} s for a nucleus to acquire its electron system and thus to form an atom or a molecule with certain chemical properties. The same requirement was mentioned in Harvey et al. (1976): "We suggest that composite nuclear systems that live less than about 10^{-14} s ... shall not be considered a new element." This view seems to have won general acceptance and has recently been repeated in Schädel (2015).

So-called quasi-atoms of very high Z are formed transiently in heavy-ion collisions, but they exist only for about 10^{-20} s. Consequently they do not qualify as nuclides of new elements. However, there seems to be no consensus among nuclear physicists of when a nucleus exists. Some take the definition of an atomic nucleus to be limited by the time scale 10^{-12} s and according to others "If a nucleus lives long compared to 10^{-22} s it should be considered a nucleus" (Thoenessen 2004, p. 1195).

The TWG realised that the phrase "beyond reasonable doubt" was vague, but the group used it deliberately to stress that such doubts could not be completely avoided. Independent confirmation of a discovery claim was of course important, as stressed by all three SHE groups, but it was not a magic wand that in all cases would eliminate doubts. "All scientific data, other than those relating to unique events such as a supernova, must be susceptible of reproduction," it was pointed out.[2] And yet, although the TWG attached much importance to reproducibility, "We do not believe that recognition of the discovery of a new element should always be held up until the experiment or its equivalent have been repeated, desirable in principle as this may be." The more relaxed attitude to the reproducibility criterion was in part practically motivated in "the immense labour and the

---

[2]For the comparison to unique but accepted astronomical events in nature, see also Armbruster and Münzenberg (2012, p. 284). There are other kinds of unique and hence non-reproducible natural phenomena, for instance magnetic storms, cosmic rays, and rare elementary particles which cannot be produced in the laboratory.

time necessary to detect even a single atom of a new element." There were circumstances, it was said, where "a repetition of the experiment would imply an unreasonable burden."

The discovery criteria of the TWG were adopted by all later working parties and thus defined for more than two decades what it meant to have discovered a new element. Only in June 2017 was a new IUPAC-IUPAP working group under the Inorganic Chemistry Division and chaired by Sigurd Hofmann established with the purpose of re-examining and updating the old criteria. According to IUPAC (2017), "The recent completion of the naming of the one-hundred and eighteen elements in the first seven periods of the Periodic Table of the elements provides a natural opportunity for a necessary expert review of these criteria in the light of the experimental and theoretical advances in the field."

The TWG evaluation of discovery claims concerning the transfermium elements with atomic numbers from 101 to 112 appeared in 1992. The aim of the detailed report was to assign priority and credit for the discoveries of the elements, not to propose names for them (Barber et al. 1992; Wilkinson et al. 1993). This was done by reviewing all relevant papers and critically comparing their results and methods with the discovery criteria and also with the most recent knowledge. In some cases the TWG conclusions were unambiguous, as they were in the case of $Z = 107$ where priority was assigned to the 1981 work of the Darmstadt group. But in other cases the conclusions were far from unambiguous. Element 103 provides an example. "Effective certainty" had been approached with papers published by the Dubna group in the late 1960s, but it was only with a Berkeley paper of 1971 that "all reasonable doubt had been expelled." So who should be credited with the discovery? The TWG tried to please both parties:

> In the complicated situation presented by element 103, with several papers of varying degrees of completeness and conviction, none conclusive, and referring to several isotopes, it is impossible to say other than that full confidence was built up over a decade with credit attaching to work in both Berkeley and Dubna.

For elements 104 and 105 the TWG similarly suggested that credit should be shared between the American and Russian scientists, something which the Berkeley scientists found most unreasonable for $Z = 104$ in particular. For $Z = 106$, on the other hand, the verdict was clear and to the satisfaction of the Americans: "The Dubna work ... does not demonstrate the formation of a new element with adequate conviction, whereas that from Berkeley-Livermore does." Priority for having discovered elements 107 to 109 was essentially, but in some cases with qualifications, assigned to the Darmstadt group. The TWG refrained from describing element 109 as discovered and merely concluded that the work of the Darmstadt scientists "gives confidence that element 109 has been observed." Proof was still lacking.

The three SHE laboratories were given the right to respond to the TWG report (TWG 1993). Oganessian and Zvára responded on behalf of Dubna, and Münzenberg on behalf of Darmstadt. While the Russian and German SHE scientists were reasonably satisfied, the Americans were not. To their eyes, the transfermium war was not quite over. Seaborg and Ghiorso wasted no time in responding by

letters to Aaldert Wapstra, secretary of the TWG, in which they expressed their
deep concern over the "seriously flawed report." On 19 November 1991, at a time
when the report had not yet been published, they wrote as follows (Hoffman et al.
2000, p. 383):

> The report's account of the discovery of element 104 is particularly disturbing to us and
> suggests the operation of a double standard in the evaluation of the work by the Berkeley
> and the Dubna groups. The Dubna identification of ^{260}104, decaying by spontaneous
> fission, is obviously wrong so an attempt is made to credit the Dubna volatility work,
> marginal and inconclusive at best, which was performed after the Berkeley discovery
> experiments. On the other hand, no mention is made in the assessment section of the first
> meaningful chemical identification of element 104 at Berkeley in 1970 by the reliable ion
> exchange method.

The two Californians similarly complained about the TWG's account of element
106.

As mentioned above, in their published response Ghiorso and Seaborg (1993)
reacted strongly against the conclusions and the TWG in general. Their "most
serious quarrel" with the report concerned element 104 and the shared credit for its
discovery to both Dubna and Berkeley. Ghiorso and Seaborg flatly denied the
validity of the TWG conclusion, which they considered "a disservice to the scientific community," and they argued that priority to element 105 did not belong
jointly to Berkeley and Dubna but to Berkeley alone. Writing ironically of "the
supposedly impartial TWG" the two Americans intimated that the TWG was in
reality pro-Dubna or had been cleverly manipulated by the Dubna group, an
opinion which also appeared in Karol (2004). The claims of the Berkeley scientists
and their proposed names rutherfordium and hahnium had received strong support
from an earlier in-depth review composed by three American scientists on which
Ghiorso and Seaborg relied (Hyde et al. 1987). But Wilkinson, the chairman of the
TWG, was not swayed by any of the American objections: "After detailed examination of all criticism from Berkeley we do not find it necessary in any way to
change the conclusions of our report" (TWG 1993, p. 1824; see also Bradley 1993).

The transfermium wars were as much about names as about priority. In the early
1990s there were several attempts to negotiate the naming problems internally
among the involved German, American and Russian nuclear scientists, but the
attempts led to no agreement. Oganessian discussed the problems with Seaborg and
other members of the Berkeley group, and in August 1992 Armbruster of GSI
proposed a compromise list of names for elements 102 to 109. The list was in part
based on the idea of separating the name of an element from its discovery, which
the Americans found unreasonable. Ghiorso and Seaborg appreciated the German
initiative but without agreeing with the proposed list of names. A planned meeting
in Paris with members of the three SHE laboratories had to be cancelled because of
American resistance (Hoffman, Ghiorso and Seaborg, p. 385; Armbruster and
Münzenberg 2012, p. 285).

Recommendations for the names of elements 101–109 were agreed upon at a
meeting of CNIC in the summer of 1994 and shortly later accepted by the IUPAC
Bureau at a meeting in Antwerp (IUPAC 1994; Greenwood 1997). The names were

chosen after the three major SHE laboratories had been consulted, but to some heavy-element scientists they came as an unwelcome surprise. The nomenclature committee of the American Chemical Society (ACS) had a few months earlier opted for rutherfordium instead of dubnium (104), hahnium instead of joliotium (105), and seaborgium instead of rutherfordium (106). Moreover, the Americans at first preferred nielsbohrium (Ns) to bohrium, for other reasons because they found the latter name to be too close to boron. For element 108 they recommended the name hassium chosen by the Darmstadt group, referring to the German state Hessen where GSI was located. To increase the confusion, the name nielsbohrium was originally proposed for element 105 by the Dubna group.

Not only did the Americans come up with names different from those of IUPAC, they also questioned the international union's authority to name elements and to disregard the proposals of the discoverers (Rothstein 1995; Rayner-Canham and Zheng 2008). However, IUPAC made it clear that it was not obliged to follow these or other proposals from the involved scientists.

Due to persistent American pressure the 1994 list of names was not officially approved by the IUPAC Council. During an IUPAC conference in Guilford, UK, in 1995 an ad hoc meeting was summoned and as a result of this meeting the 1994 list was replaced by a different list of names. The new list was thought, mistakenly, to be a Solomonic compromise acceptable to all parties. This time nobelium was replaced with flerovium for element 102, and joliotium was suggested for element 105. But the revised list of names was no more successful than the previous one and never won official recognition. According to Darleane Hoffman (1996), the revision "created a chaotic situation in heavy element nomenclature."

Only in 1997 was the final decision taken by the IUPAC Council at the General Assembly in Geneva. Dubnium was still on the list but now for element 105, flerovium had disappeared and so had joliotium (IUPAC 1997; Fontani et al. 2015, pp. 366–388). While the discoverers of element 107 wanted to call it nielsbohrium, after consultations with the Danish IUPAC committee it was decided to retain the former bohrium. The Americans, including the ACS, generally approved the new slate of names but continued to have reservations with regard element 105. The reason why the ACS Committee on Nomenclature nonetheless accepted the name is worth quoting: "In the interest of international harmony, the Committee reluctantly accepted the name 'dubnium' for element 105 in place of 'hahnium,' which has had long-standing use in literature" (quoted in Hoffman et al. 2000, p. 395).

One name that did not make it to the official lists, but was advocated with much vigour by the Dubna scientists, was kurchatovium for element 104 (and at one stage also for element 106). The name was originally proposed by Flerov and his group in honour of Igor Kurchatov, a nuclear physicist known as the father of the Soviet atomic bomb, and it was strongly supported by Oganessian. For more than two decades kurchatovium (chemical symbol Ku) appeared regularly in scientific journals from the Soviet Union and its allied countries. The third edition of The Great Soviet Encyclopedia contained a detailed entry under the name written by Flerov (1979). Although kurchatovium remained unrecognised in the West, it was positively received by some scientists of a communist inclination. "I remember how, in 1970, some chemists at the University of Bordeaux, France, who happened to be fashionably left-wing and pro-Soviet, insisted that element number 104 was to be called *kourtchatovium*" (Diament 1991, p. 209).

Seaborg was for a time willing to consider kurchatovium for element 106, but only as a trade-off with the Dubna group if it admitted the Berkeley claim to the discovery of element 104. Yet, in a letter to Armbruster from the summer of 1992, Ghiorso and Seaborg made it clear that the name was unacceptable (Hoffman et al. 2000, p. 385): "We would never agree to the naming of an element after Kurchatov (anymore than we would to the naming of an element after an American inventor of the hydrogen bomb)—and certainly not to the naming of element 105 after Kurchatov!" This was also the message from Michael Nitschke, a German-born member of the Berkeley team (Armbruster and Münzenberg 2012, p. 285) (Table 5.1):

> Jointly with our Dubna colleagues we would gladly suggest a name for element 104 in honor of a great Russian scientist with international reputation, as we did in honouring Mendelejev in the case of element 101. We cannot accept Kurchatov as meeting this criterion. If this suggestion is unacceptable, it is my impression that the negotiations have failed. In this case no American delegation will come to Paris.

5.3   The Case of Seaborgium

According to the TWG reports of 1992 and 1993 the experiments made by the Berkeley-Livermore group in 1974 demonstrated the existence of element 106 "with adequate conviction." The nearly simultaneous work of the Dubna group was recognised to be important but not carrying quite the same conviction. While priority to the discovery of the element was thus assigned the American scientists, in its report of 1994 CNIC recommended a name for it (rutherfordium) which disregarded the proposals of the discoverers. Neither the Berkeley-Livermore group nor the rival Dubna group had suggested a name for the new element in their papers of 1974. Members of the two groups had met in Berkeley and exchanged information about their experiments, and they agreed that names should be postponed until their observations had been confirmed (Ghiorso et al. 1974, p. 1493).

In the fall of 1994 a Berkeley team reported a complete confirmation of the original discovery claim, thus "permitting the discoverers to propose a name for element 106" (Gregorich et al. 1994). Several names were proposed and discussed in Berkeley, some of them seriously and others more jokingly. At one point Ghiorso favoured "alvarezium" in honour of the Berkeley physicist Luis Alvarez, a Nobel Prize laureate who had passed away in 1988 (Browne 1993). But in the end Ghiorso and his team settled on "seaborgium" and after some hesitation the 82-year-old veteran in transuranic science assented (Seaborg 1995). The proposal was announced at a meeting of the ACS in March 1994 and the name unanimously endorsed by the ACS Committee on Nomenclature. Contrary to other cases of SHE

Table 5.1 Names for elements 102–109 proposed between 1992 and 1997

Z	GSI 1992	ACS 1994	IUPAC 1994	IUPAC 1995	IUPAC 1997
102	joliotium	nobelium	nobelium	flerovium	nobelium, No
103	lawrencium	lawrencium	lawrencium	lawrencium	lawrencium, Lr
104	meitnerium	rutherfordium	dubnium	dubnium	rutherfordium, Rf
105	kurchatovium	hahnium	joliotium	joliotium	dubnium, Db
106	rutherfordium	seaborgium	rutherfordium	seaborgium	seaborgium, Sg
107	nielsbohrium	nielsbohrium	bohrium	nielsbohrium	bohrium, Bh
108	hassium	hassium	hahnium	hahnium	hassium, Hs
109	hahnium	meitnerium	meitnerium	meitnerium	meitnerium, Mt

discoveries, the Dubna scientists did not object and they did not propose an alternative name for the element. The decision to favour rutherfordium over seaborgium was thus IUPAC's alone.

When the CNIC recommendation became known half a year later, it met with a storm of opposition not only from the powerful ACS and from the American scientific community generally but also from the international heavy-ion community. Ghiorso found the recommendation to be simply "outrageous" (Browne 1994). Charles Shank, the director of the Lawrence Berkeley National Laboratory, strongly defended seaborgium and the discoverers' privilege to name an element (Yarris 1994). And Paul Karol deplored the "madness" of CNIC, which had rejected "for the first time in history, the name picked up by the undisputed discoverers of an element because the person so honoured was still alive." He had no confidence at all in what he called "the ongoing IUPAC transfermium fiasco" and which he later characterised as a "soap opera" (Karol 1994; Karol 2004; Lehrman 1994).

Central to the vigorous but short-lived dispute were two questions: (i) Do discoverers have the right to name an element? (ii) Can an element be named after a living person? With regard to the first question, IUPAC insisted that although discoverers have a right to *suggest* a name, it is IUPAC alone which makes the decision. As to the second question, in August 1994 the twenty members of CNIC approved with 16 to 4 votes that "an element should not be named after a living person" (IUPAC 1994, p. 2420). Although five of the CNIC members were Americans, the vote against seaborgium was passed with 18 to 2. As one of the Americans supporting the majority view repeated, "Discoverers don't have the right to name an element; they have a right to suggest a name. And, of course, we didn't infringe on that at all" (Yarris 1994). Not only was there no precedence for naming elements after living scientists, the commission also argued that it was necessary to have a proper historical perspective in relation to the discoveries of elements before the decision of a name could be made.

Seaborg protested: "In the case of element 106, this would be the first time in history that the acknowledged and uncontested discoverers of an element are denied

the privilege of naming it" (Rayner-Canham and Zheng 2008, p. 17). In an interview of April 1995 he admitted to be disappointed. "The reasons they give just
don't make any sense," Seaborg said (Hargittai 2003, pp. 2–17). "They say that
they don't want to name an element after a living person … [and] also say that they
want a perspective on history."

Supporters of the Berkeley-ACS position argued that there were in fact historical
precedents, namely in the cases of einsteinium and fermium. As the Berkeley
scientists pointed out, they had used the name einsteinium in notebooks while
Einstein was still alive. However, the argument bore little weight since the papers in
which the names were proposed were published after the deaths of Fermi and
Einstein on 28 November 1954 and 18 April 1955, respectively, and the names
approved by IUPAC only in 1957 (Koppenol 2005). American scientists published
several short papers in *Physical Review* in 1954 on the new elements 99 and 100,
but at the time without suggesting names. The discovery paper by Ghiorso and
collaborators in the August issue of *Physical Review* was received on 20 June, two
months after Einstein's death. At about the same time they announced the names
einsteinium and fermium at the Geneva International Conference on the Peaceful
Use of Atomic Energy taking place in August 1955.

It is possible, as suggested in Seaborg (1994), that the Ghiorso group "decided
on the names … while Albert Einstein and Enrico Fermi were still alive," but if so it
was informally only and hence, according to IUPAC, of no relevance. A few
months after Fermi's death and shortly after Einstein's, Ghiorso informed Laura
Fermi, the widow of Enrico Fermi, about the group's discussion of a name for
element 100:

> I thought you might like to might like to know that we are planning to name element 100 in
> honor of Enrico…. It was my fortune and privilege to know your husband in the days of the
> Metallurgical Laboratory project, and I can say from personal contact that science has lost a
> very warm-hearted human being as well as its greatest physicist.[3]

The letter suggests that still in April 1955 the Ghiorso group had not made a final
decision about the name.

Even less convincing is the claim that the name gallium for element 31
(eka-aluminium) was proposed by and named after its discoverer, the French
amateur chemist Emile Lecoq de Boisbaudran. According to Seaborg, "gallium was
named after a living person in 1875" (Hargittai 2003, p. 5). The claim, based upon
the French name for *gallus* (cock) being *coq*, was later repeated and elaborated in
Karol (2004, p. 256). However, Lecoq de Boisbaudran always maintained that he
had named the element after his fatherland France (Fontani et al. 2015, p. 169). As
The Economist generalized in a comment to the naming controversy, "When it
comes to giving things names, scientists have a habit of throwing logic and consistency out of the window" (Economist 1998).

---

[3]Letter of 26 April 1955, as quoted by Seaborg in a symposium of 1978 commemorating the
twenty-fifth's anniversary of elements 99 and 100. Online as http://escholarship.org/uc/item/
92g2p7cd.

The IUPAC 1994 recommendation of names met opposition not only from American nuclear scientists but also from some of their colleagues from Russia and Germany. In a letter sent to IUPAC in March 1995 three leading nuclear scientists, one from each of the countries, emphasised that the discoverers of a new element should also have the right to name it. According to the three scientists, V. Goldanski (Russia), D. Hoffman (USA) and J. Kratz (Germany): "There should be no new retroactive rule that elements cannot be named after living persons" (Karol 2004; Hoffman et al. 2000, p. 391). IUPAC could have maintained its position, but in the end it did not. At a meeting in late August 1996 CNIC "decided to modify its decision that the name of a living scientist should not be used as the basis for an element name" (IUPAC 1997, p. 2472). The modification was ratified by the IUPAC Council a year later.

Seaborgium was the first element ever named after a scientist during his lifetime. Only in 2016, with the approval of oganesson for element 118, was another element named after a living scientist, in this case after Yuri Oganessian, another octogenerian. While many American scientists welcomed the change in IUPAC's naming policy, not all did. Allan Bromley, a former president of IUPAP, regretted that the original recommendation had been overturned as a result of strong political pressure, which he found was "unworthy of the scientific community."[4]

The question of the names of artificially produced elements was not a new one. In 1947, at a time when eight such elements were known (in addition to the transuranic elements also $Z = 43, 61, 85$, and 87), the eminent Austrian-British radiochemist Friedrich Paneth addressed the question. He suggested three rules that he thought would lead to greater unity and consistency in the chemical names (Paneth 1947, p. 8; Koppenol 2005):

(1) The right to name an element should go to the first to give a definite proof of the existence of one of its isotopes. (2) In deciding the priority of the discovery, there should be no discrimination between naturally occurring and artificially produced isotopes. (3) If a claim to such a discovery has been accepted in the past, but is refuted by later research, the name given should be deleted and replaced by one chosen by the real discoverer.

The second of the suggestions was at the time uncontroversial, but the first one was not and was in fact rejected by IUPAC at its London conference the same year. Nor did IUPAC or the majority of scientists accept the third suggestion, which too easily might lead to confusion and frequent changes of names. Still, during the period of the transfermium wars this is what happened with several of the names assigned to the superheavy elements. The procedure suggested by Paneth was not ineffective, though, as the revision of element names adopted by IUPAC in 1949 was directly inspired by Paneth's rules (Coryell and Sugarman 1950).

---

[4]Bromley to Kumar, 5 January 1998. In http://www.marinov-she-research.com/image/users/352030/ftp/my_files/External/Bromley%20Letter%201998.pdf.

References

Armbruster, P.: On the production of heavy elements by cold fusion. The elements 106 to 109. Annu. Rev. Nucl. Part. Sci. **35**, 135–194 (1985)

Armbruster, P., Münzenberg, G.: An experimental paradigm opening up the world of superheavy elements. Eur. Phys. J. H **37**, 310–327 (2012)

Barber, R.C., et al.: Discovery of the transfermium elements. Prog. Part. Nucl. Phys. **29**, 453–530 (1992)

Bradley, D.: Battle resumes over who found heavy elements. New Scientist **139**, 8–9, 14 August 1993

Browne, M. V.: Advance made in seeking heavy elements. New York Times, 12 October 1993

Browne, M. W.: Element is stripped of its namesake. New York Times, 11 October 1994

Coryell, C.D., Sugarman, N.: The acceptance of new official names for the elements. J. Chem. Educ. **27**, 460–461 (1950)

Diament, H.: Politics and nationalism in the naming of chemical elements. Names: J. Onomast. **39**, 203–216 (1991)

Economist.: Today we have naming of parts. The Economist. https://www.astro.com/swisseph/econ4686.htm December 1998

Fennell, R.: History of IUPAC 1919–1987. Blackwell Science, London (1994)

Fernelius, W.C., Loening, K., Adams, R.M.: How are elements named? J. Chem. Educ. **48**, 730–731 (1971)

Flerov, G.N.: Synthesis and search for heavy transuranium elements. Sov. Atom. Energy **28**, 390–397 (1970a)

Flerov, G.N.: Soviet synthesis of element 105. Science **170**, 15 (1970b)

Flerov, G.N.: Kurchatovium. The Great Soviet Encyclopedia. https://encyclopedia2.thefreedictionary.com/kurchatovium (1979)

Flerov, G.N., et al.: History of the transfermium elements $Z = 101, 102, 103$. Sov. J. Part. Nucl. **22**, 453–483 (1991)

Fontani, M., Costa, M., Orna, M.V.: The Lost Elements: The Periodic Table's Shadow Side. Oxford University Press, Oxford (2015)

Ghiorso, A.: Disputed discovery of element 105. Science **171**, 127 (1971)

Ghiorso, A., et al.: Element 106. Phys. Rev. **33**, 1490–1493 (1974)

Ghiorso, A., Seaborg, G.T.: Response to the report of the Transfermium Working Group 'Discovery of the transfermium elements'. Prog. Part. Nucl. Phys. **31**, 233–237 (1993)

Greenwood, N.N.: Recent developments concerning the discovery of elements 101-111. Pure Appl. Chem. **69**, 179–184 (1997)

Gregorich, K.E., et al.: First confirmation of the discovery of element 106. Phys. Rev. Lett. **72**, 1423–1426 (1994)

Hargittai, I.: Candid Science III: More Conversations with Famous Chemists. Imperial College Press, London (2003)

Harvey, B.G., et al.: Criteria for discovery of chemical elements. Science **193**, 1271–1272 (1976)

Hoffman, D.C.: the transuranium elements: From neptunium and plutonium to element 112. In: Unpublished Conference Paper. http://www.iaea.org/inis/collection/NCLCollectionStore/_Public/28/017/28017156.pdf (1996)

Hoffman, D.C., Ghiorso, A., Seaborg, G.T.: Transuranium People: The Inside Story. Imperial College Press, London (2000)

Holcomb, R.W.: Element 105 synthesized and named hahnium by Berkeley researchers. Science **168**, 810 (1970)

Huizenga, J.R.: Cold Fusion: The Scientific Fiasco of the Century. University of Rochester Press, Rochester, NY (1992)

Hyde, E.K., Hoffman, D.C., Keller, O.L.: A history and analysis of the discovery of elements 104 and 105. Radiochim. Acta **42**, 57–102 (1987)

IUPAC: Names and symbols of transfermium elements. Pure Appl. Chem. **66**, 2419–2421 (1994)

IUPAC: Names and symbols of transfermium elements. Pure Appl. Chem. **69**, 2471–2473 (1997)

IUPAC.: Joint working group to examine the 1991 criteria used to verify the discovery of new elements. https://iupac.org/projects/project-details/?project_nr=2017-014-2-200 (2017)

Karol, P.J.: Transfermium wars. Chem. Eng. News **18**, 2–3, 31 October 1994

Karol, P.J.: The heavy elements. In: Ouvray, D.H., King, R.B. (eds.) The Periodic Table: Into the 21st Century, pp. 235–262. Research Studies Press, Baldock, England (2004)

Koppenol, W.H.: Paneth, IUPAC, and the naming of elements. Helv. Chim. Acta **88**, 95–99 (2005)

Lehrman, S.: 'Seaborgium' fails to win approval. Nature **371**, 639 (1994)

Oganessian, Y.T., Lazarev, Y.A.: Problems involved in the synthesis of new elements. Pure Appl. Chem. **53**, 925–947 (1981)

Paneth, F.A.: The making of the missing chemical elements. Nature **159**, 8–10 (1947)

Rayner-Canham, G., Zheng, Z.: Naming elements after scientists: an account of a controversy. Found. Chem. **10**, 13–18 (2008)

Rothstein, L.: The transfermium wars. Bull. At. Sci. **51**, 5–6 (1995) (January)

Schädel, M.: Chemistry of the superheavy elements. Phil. Trans. R. Soc. A **373**, 20140191 (2015)

Seaborg, G.T.: Terminology of the transuranium elements. Terminology **1**, 229–252 (1994)

Seaborg, G.T.: Transuranium elements: past, present, and future. Acc. Chem. Res. **28**, 257–264 (1995)

Seaborg, G.T., Loveland, W.D.: The Elements Beyond Uranium. Wiley, New York (1990)

Thoenessen, M.: Reaching the limits of nuclear stability. Rep. Prog. Phys. **67**, 1187–1232 (2004)

Thompson, S.G., Tsang, C.F.: Superheavy elements. Science **178**, 1047–1055 (1972)

TWG.: Responses on the report 'discovery of the transfermium elements'. Pure Appl. Chem. **65**, 1815–1824 (1993)

Wapstra, A.H.: Criteria that must be satisfied for the discovery of a new chemical element to be recognized. Pure Appl. Chem. **63**, 879–886 (1991)

Wilkinson, D.H., et al.: Discovery of the transfermium elements. Pure Appl. Chem. **67**, 1757–1814 (1993)

Yarris, L.: Naming of element 106 disputed by international committee. http://www.lbl.gov/Science-Articles/Archive/seaborgium-dispute.html (1994)

Zvára, I.: Studies of the heaviest elements at Dubna. Dubna preprint, see http://www.iaea.org/inis/collection/NCCLCollectionStore/_Public/06/179/6179997.pdf (1973)

Chapter 6
Super-Superheavy Elements

Abstract Since the mid-1990s nine more superheavy elements synthesised in either cold or hot fusion processes have entered the periodic table. Elements with atomic numbers 110–112 were produced in Darmstadt, Germany, whereas most of the other elements owed their discoveries to collaborations of Russian and American scientists. The exception to the German-Russian-American hegemony was element 113 discovered by a team of Japanese scientists and named nihonium after their country. The heaviest of all elements so far, oganesson with $Z = 118$, is presently the only element named after a still living scientist. Apart from outlining the discovery histories of the very heavy elements the chapter also considers the role played by IUPAC and associated working groups in the formal and final recognition of the new elements. It ends with references to some quite unserious aspects of hypothetical superheavy elements and their often fanciful names.

Keywords Superheavy elements · IUPAC · Joint working party
Yuri oganessian · Kōsuke morita · Heaviest element

At present the periodic table is complete with known elements up to and including element 118. The chapter outlines the discovery histories of the heaviest and most recently recognised elements and discusses how they were assigned their names. Numerous other names have been proposed, some of them quite funny and others just silly. They belong to the lighter side of the science of heavy elements.

6.1   The Heaviest Elements So Far

The elements with atomic numbers 110, 111 and 112 were produced, almost (but not quite) routinely, by the Darmstadt GSI scientists in the brief period 1994–1996 (Hofmann 2002, pp. 163–174; Armbruster and Münzenberg 2012, pp. 288–300). A few atoms of the first element, namely darmstadtium isotopes of mass numbers

269 and 271, were produced in late 1994 in reactions where nickel ions were smashed into a Pb-208 target:

$$^{208}_{82}\text{Pb} + ^{62}_{28}\text{Ni} \rightarrow ^{269}_{110}\text{Ds} + ^{1}_{0}n \quad \text{and} \quad ^{208}_{82}\text{Pb} + ^{64}_{28}\text{Ni} \rightarrow ^{271}_{110}\text{Ds} + ^{1}_{0}n$$

There had been earlier attempts to synthesise the element at the laboratories in Dubna, Berkeley and Darmstadt, but none of these did conclusively establish its existence. Although the Darmstadt group faced competition from the Dubna scientists in particular, the relationship between the two groups was friendly and cooperative. In fact, the Darmstadt team behind the discovery of the three elements included three scientists from Dubna. Since the early 1980s the German group also had close connections to American researchers which included an exchange of scientists between GSI and LBNL (Lawrence Berkeley National Laboratory). The Darmstadt group generally mediated between the American and Russian rivals.

The first event revealing the existence of element 110, darmstadtium, was observed on 13 November 1994. Sigurd Hofmann (2004, pp. 166–168) tells what happened next:

> At about noon that very day it was clear that we had to publish our result as soon as possible because of the experiment running at Dubna. ...We agreed not to delay by telling anybody about the discovery and to finish the draft paper by that evening. ...Next morning, November 14th, the final version of the paper was finished and one of our drivers ... delivered it directly to the editors [of *Zeitschrift für Physik*]. We also officially informed the GSI staff and a press release was issued on the 17th; the following day we read about our results in the newspapers.

The next element, to be named roentgenium, was produced a few weeks after element 110 was first observed. Two of the co-discoverers later described the discovery of element 111 as just "an easy step further" (Armbruster and Münzenberg 2012, p. 293). Also in this case nickel ions were used as projectiles, but now with bismuth as target material:

$$^{209}_{83}\text{Bi} + ^{64}_{28}\text{Ni} \rightarrow ^{272}_{111}\text{Rg} + ^{1}_{0}n$$

In the first GSI experiment just a single atom was detected and in later experiments of 2002 three more were found. The close connection between the discoveries of elements 110 and 111 is illustrated by their appearance in the form of companion papers in volume 350 of *Zeitschrift für Physik*.

The third product of the Darmstadt offensive came forth in early 1996 when an atom of element 112 was detected in reactions between Pb-208 and Zn-70 to be repeated four years later. In an earlier section (Sect. 4.1) we have dealt with the remarkable history of this element first claimed discovered in 1971 and we have also referred to the proposed name, copernicium, and the change of its chemical symbol from Cp to Cn. In all three cases the discoveries were announced in the German but at the time English-language *Zeitschrift für Physik*, once the world's most prestigious journal in atomic and quantum physics but at the time hardly a leading journal in heavy-ion physics and SHE research. In 1998 the venerable journal ceased to exist as it merged with the French *Journal de Physique* into the new *European Physical Journal*.

The first two elements received official recognition relatively quickly—or by IUPAC standards very quickly. "Element 110 has been discovered by this collaboration" stated the JWP report of 2001, referring to the 1995 paper of the Darmstadt group. On the other hand, the JWP concluded that the Darmstadt claims for elements 111 and 112 were still not sufficiently confirmed. German experiments of 2001 were made to certify the previously measured data and thus to comply with the JWP requirements. Hofmann et al. (2002) cited the criteria of the 1992 TWG report and briefly discussed them in relation to the Darmstadt experiments. Having provided the necessary confirmation for element 111, the JWP report of the following year accepted the discovery claim (Karol et al. 2003). The last of the trio, element 112, had to wait until 2009 before the JWP was satisfied and accepted the Darmstadt claim of 1996 as "the first evidence for the synthesis of element with atomic number 112" (Barber et al. 2009). An important reason for the changed evaluation was confirming experiments made by physicists at the RIKEN heavy-ion facility in Japan.

By and large, elements 114 and 116 were Dubna discoveries made by a team led by Yuri Oganessian and with participation of American researchers from LLNL, the Lawrence Livermore National Laboratory. The transfermium war was over. The early history of $Z = 114$ included several unconfirmed and doubtful events ascribed to decay of just one or two atomic nuclei. In an experiment from December 1998 the Dubna team observed an alpha decay chain from which they inferred that the parent nucleus was most likely produced in the reaction

$$^{244}_{94}\text{Pu} + {}^{48}_{20}\text{Ca} \rightarrow {}^{290}_{114}\text{Fl} + 2{}^{1}_{0}n$$

The claimed observation of this isotope and also of Fl-287 could not be verified and it was only in later experiments that a Dubna-LLNL team succeeded in confirming the creation of atoms of element 114. With the further confirmation of isotopes of mass numbers between 286 and 289 by experiments made in Berkeley and Darmstadt, there was no doubt that the element had been discovered.

After the false discovery claim of 1999 (Sect. 4.2) attempts to produce element 116 were continued by the Dubna and Livermore scientists who in experiments of 2000 bombarded a Cm-248 target with doubly magic Ca-48 ions. "On the 35th day of irradiation, after an accumulated beam dose of 6.6×10^{18} ions, the first event sequence was observed that can be assigned to the implantation and decay of the isotope of element 116 with mass number 292" (Oganessian et al. 2000). The mass number of the lonely nuclide was later altered to 293 and in subsequent experiments a handful of more atoms of element 116 turned up. In its report of 2011 the JWP judged that whereas the early work on elements 114 and 116 was inconclusive, the later experiments proved that the two elements had now been discovered beyond any reasonable doubt: "The JWP … recommends that the Dubna-Livermore collaboration be credited with the discovery of this new element [$Z = 114$]. In a similar manner, … the Dubna-Livermore collaboration should be credited with the

Table 6.1 Names and discoveries of elements 110–118 recognised by IUPAC

Z	Name, symbol	Discovery	Group	JWG/JWP recognition	IUPAC recognition
110	darmstadtium, Ds	1995	Darmstadt	2001	2003
111	roentgenium, Rg	1995	Darmstadt	2001	2004
112	copernicium, Cn	1996	Darmstadt	2009	2009
113	nihonium, Nh	2004	RIKEN	2016	2016
114	flerovium, Fl	1999	Dubna-Livermore	2011	2012
115	moscovium, Mc	2010	Dubna-Livermore-ORNL	2016	2016
116	livermorium, Lv	2004	Dubna-Livermore	2011	2012
117	tennessine, Ts	2010	Dubna-Livermore-ORNL	2016	2016
118	oganesson, Og	2006	Dubna-Livermore	2016	2016

discovery of the new element with $Z = 116$" (Barber et al. 2011, p. 1494). The discovery and recognition history of elements 110–118 is summarized in Table 6.1.

The most recent newcomers to the periodic table are the elements with atomic numbers 113, 115, 117 and 118, which all received official recognition in 2016. The announcement took form of a press release of 30 December 2015 which apparently was due to IUPAC alone, without consulting IUPAP (IUPAC 2015; Jarlskog 2016).

Disregarding contemporaneous discovery claims by the Dubna-Livermore collaboration, the first atoms of element 113 were produced in 2003–2005 by a team led by Kōsuke Morita at the RIKEN Nishina Center for Accelerator-Based Science (RNC) in Japan. The name refers to Yoshia Nishina, a pioneer Japanese physicist and former collaborator of Niels Bohr. Interestingly, it is possible that Nishina in the mid-1920s, on the instigation of Bohr, searched for transuranic elements in uranium (Kragh 2013). While previously SHE synthesis and research had been essentially limited to institutions in the United States, Russia/USSR and Germany, since the 1990s the RIKEN group entered the game as an important fourth player. Morita and his collaborators achieved their final results in 2012, published in the *Journal of the Physical Society of Japan* and using the cold fusion method to produce an atom of Nh-278. For an atom as heavy as the one of element 113 the reaction probability of the synthesis process was very low, requiring the Bi target to be irradiated with Zn-70 projectiles for no less than 450 days. "In a sense we are doing what the alchemists were trying to do," Morita commented to the *Asian Scientist* magazine in 2016.

The discovery histories of elements 115 and 117 were closely intertwined. In experiments from 2004 Oganessian's Dubna-Livermore collaboration not only claimed the discovery of $Z = 113$ but also of $Z = 115$ as the alpha parent element of the former. Using Am-243 as a target and the standard Ca-48 as projectile, the group found it "most reasonable" that the observed alpha decay chains "originate from parent isotopes of element 115." The isotopes were produced in reactions

$$^{243}_{95}\text{Am} + ^{48}_{20}\text{Ca} \rightarrow ^{291-x}_{115}\text{Mc} + x^1_0 n$$

where $x = 3$ or 4 (Oganessian et al. 2004). Later experiments resulting in isotopes of element 115 were confirmed by scientists at Darmstadt, Berkeley and Lund in Sweden. The simultaneous experiments leading to the discovery claim of element 117 in 2010 involved not only the Dubna and Livermore groups but also scientists from the Oak Ridge National Laboratory (ORNL) in Tennessee. As illustrated by the discovery paper in which element 117 was announced, SHE production was big science from a nuclear physics point of view (if not from the point of view of high energy physics). The paper by Oganessian and his collaborators (2010) included 33 authors from six different institutions.

Although the JWP did not accept the RIKEN claim for element 113 as a proved discovery in its 2011 report, after further work it was recognised together with elements 115 and 117. Priority to element 113 was credited the RIKEN group, and after further experiments and confirmations from other groups the JWP stated in its 2016 report that "[the] collaborations of Oganessian et al. have met the Criteria for discovery of the elements with atomic numbers $Z = 115$ and $Z = 117$" (Karol et al. 2016). Recalling the past history, the working party cautiously added: "Should the recommendations of the JWP prove, though future experiments, to be subject to reversal, there should be no issue with authorizing revisions as this has occurred in the past, viz with nobelium."

And finally to element 118 which was another product of the successful Dubna-Livermore collaboration. Its discovery history was summarized in Sect. 4.2 and here we just note that so far no more than a handful of nuclei have been produced. The only known isotope is Og-294, which makes it the element with the fewest isotopes currently known. The nuclide decays with a lifetime less than 1 ms. None of the physical and chemical properties of the element have been measured. They are all calculated or extrapolated, including the number and configuration of the element's atomic electrons. Of course, no chemical compounds of oganesson are known (see also Sect. 7.3).

6.2 Names

Generally the names of the heaviest of the elements were agreed upon un-dramatically and without the heated controversies which had characterised some of the earlier assignments of names. As Seaborg wrote in a letter of 3 November 1997 to the chairman of CNIC, he had no problem with leaving the names of elements 111 and 112 to his colleagues in Darmstadt. However, he and his Berkeley group advocated that the name "hahnium," which had been previously used for element 105 and also for element 108, was now transferred to element 110

(Hoffman et al. 2000, p. 397). This did not happen and to this day no element has been officially named after Otto Hahn.

Names for the elements 110–112 discovered in Darmstadt were discussed by the GSI team at a "names-finding day" in December 1997 resulting in a preliminary list of 30 names (Hofmann 2002, pp. 176–181). After the three elements had won IUPAC recognition the names darmstadtium, roentgenium and copernicium were proposed and they were accepted by IUPAC in 2003, 2004 and 2010, respectively. Nihonium, the name for element 113, the first one ever discovered in Asia and by scientists from this continent, was officially proposed in 2016 and approved by IUPAC the same year (Öhrström and Reedijk 2016). The name derives from the country where it was discovered, "Nihon" literally meaning "the land of the rising Sun." At an earlier date various other names had been considered, such as japonium, rikenium and nishinamium, the latter a reference to Yoshio Nishina.

The names of elements 114 and 116 discovered by the Dubna-Livermore collaboration did not involve any significant controversy either. On the suggestion of the Dubna team the first element was named flerovium, not after Georgii Flerov personally but after the Flerov Laboratory of Nuclear Reactions, a part of the JINR. The same kind of institutional naming was used for element 116, livermorium, the name being a reference to the Lawrence Livermore National Laboratory in California. Both names were approved by the IUPAC Bureau in 2012 (Loss and Corish 2012). With livermorium the state of California had been honoured with the names of three elements, the two others being californium and berkelium. Impressive as it is, it is nothing compared to the Swedish village Ytterby after which four elements have been named (yttrium, terbium, erbium, and ytterbium).

Scientists from the Dubna-Livermore-ORNL collaboration which had discovered elements 115 and 117, agreed diplomatically to share the names between a Russian and an American locality. Element 115 was named moscovium after Moscow, and element 117 was called tennessine after the state of Tennessee where the Oak Ridge laboratory is located. The proposals were accepted by IUPAC in 2016.

Element 113 was named nihonium at the same time that names for elements 115, 117 and 118 were officially assigned. Following a five-month period of public review, the names recommended by the Inorganic Chemistry Division under IUPAC were announced on 8 June 2016 and finally approved by the IUPAC Bureau on 28 November 2016 (IUPAC 2016). They were ratified by the IUPAC Council at a meeting in July 2017. The whole process caused public attention and there were several speculations of what names would be assigned to the new quartet of elements, "moseleyon" for $Z = 118$ being one of the suggestions (Burdette et al. 2016). In March 2016 scientists of the tripartite Russian-American collaboration agreed to name the heaviest of the elements oganesson in honour of the Russian luminary in SHE research (Chapman 2016). This was only the second time in the history of chemistry that an element was named after a living scientist, the first instance being element 106 named after Glenn Seaborg (Sect. 5.3).

Names and naming procedures for new elements have been considered in Sects. 3.1, 4.1, 5.2, and 5.3. According to the IUPAC report of 2002, after the JWP has accepted a discovery claim, the union's Inorganic Chemistry Division will invite the discoverers to propose a name and a symbol. If the Inorganic Chemistry Division has no objections, a provisional recommendation will be reviewed by experts and interested parties during a period of five months after which the rec-ommendation is forwarded to the IUPAC Council for final acceptance and formal naming (Koppenol 2002). This procedure remained essentially unaltered except that it was slightly revised to accommodate the discoveries of elements 117 and 118. The earlier report stated that for "linguistic consistency" all elements should end in "-ium," but the revised version recommended "-ine" for elements belonging to group 17 and "-on" for elements of group 18 (Koppenol 2016; Corish 2016). Hence tennessine and oganesson.

6.3 The Lighter Side of the Heaviest Elements

The discoveries of the last quartet of SHEs revived scientific as well as popular speculations about an end to the periodic system (Hadhazy 2015). There would undoubtedly be discoveries of elements with $Z > 118$ in the future, but how many? Would there be a limit? The old prediction of $Z_{max} \cong 137$ reappeared in new dressings, in some cases lending the authority of the famous physicist Richard Feynman who supposedly had calculated the limit from relativistic quantum mechanics. The more than exotic element "feynmanium" appears on the internet, but not in scientific contexts. I have only found a single reference to Feynman's alleged prediction (Burdette et al. 2016) and by inspection of Feynman's cited paper in *Physical Review* of 1948 it turns out that it has nothing to do with chemical elements or the periodic table. "Feynmanium" is a joke, untriseptium being one more hypothetical element one can safely forget about.

Another element in the same dubious category is the even more exotic "ad-ministratium" announced by an anonymous chemist in 1990. The satire will be readily recognised and appreciated (Anon. 1990):

> The heaviest element known to science was recently discovered by University physicists. The element, tentatively named Administratium (AD), has no protons or electrons, which means that its atomic number is 0. However, it does have 1 neutron, 125 assistants to the neutron, 75 vice-neutrons, and 11 assistants to the vice-neutrons. This gives it an atomic mass number of 312. ... Since it has no electrons, it is inert. However, it can be detected chemically because it seems to impede every reaction in which it is present. According to one of the discoverers of the element, a small amount of Administratium made one reaction that normally takes a second take over four days. ... Research at other laboratories seems to indicate that Administratium might occur naturally in the atmosphere. According to one scientist, Administratium is most likely to be found on college campuses, in large corpo-rations and at government centers, near the best-appointed and best-maintained buildings.

This was in 1990. Since then the amount of atmospheric administratium has only increased and is today possibly greater than the amount of oxygen and silicon in the Earth's crust. Numerous other names have been proposed for the very heavy elements, some seriously and others less so. Examples are atlantisium, vulcanium, cosmium, russium, futurium, bastardium, cyclotronium, losalium, hawkingium, oberon, schrodingerine, sexium, yorkium, mechanicium, and eternium. And there are many more.

The heaviest of all superheavy elements? Well, as far as fantasies are concerned, the element of atomic number 10^{21} probably holds the world record. It has been suggested (with tongue in cheek, of course) to call the element for "godzillium" (Karol 2002). And yet godzillium is peanuts compared to the truly ultimate atom which pioneer cosmologist Georges Lemaître proposed in 1931. In this first big-bang hypothesis ever he conceived the original universe as one huge atomic nucleus "the atomic weight of which is the total mass of the universe." The corresponding atomic number comes out as $Z \sim 10^{78}$. A few years later the American astronomer Paul Merrill—who would later be the first to detect technetium in nature —referred to Lemaître's explanation of our chemical elements as descendants of much heavier elements. "Perhaps," Merrill (1933) said,

> ... we are already too late for some of the original heavier elements, but just in time for uranium, thorium and radium which will, in turn, soon be exhausted. Future chemists may speculate about them just as we speculate about elements heavier than uranium. ... Carried to its logical limit the theory postulates an original universe in the form of one immense super-radioactive cosmic atom. It is a daring speculation, but a beautiful and suggestive one.

Lemaître (1931) did not think of his primeval atom as a chemical element in the ordinary sense and he wisely avoided suggesting a name for it. But perhaps "universium" would be appropriate.

References

Anon: New element discovered. SEAC Commun. **8**, 3 (1990)

Armbruster, P., Münzenberg, G.: An experimental paradigm opening up the world of superheavy elements. Eur. Phys. J. H **37**, 310–327 (2012)

Barber, R.C., et al.: Discovery of the element with atomic number 112. Pure Appl. Chem. **81**, 1331–1343 (2009)

Barber, R.C., et al.: Discovery of the elements with atomic numbers greater than or equal to 113. Pure Appl. Chem. **83**, 1485–1498 (2011)

Burdette, S.C., et al.: Another four bricks in the wall. Nature Chem. **8**, 283–288 (2016)

Chapman, K.: What it takes to make a new element. Chem. World. https://www.chemistryworld. com/what-it-takes-to-make-a-new-element/1017677.article (2016)

Corish, J.: Procedures for the naming of a new element. Chem. Int. **38**(March), 9–11 (2016)

Hadhazy, A.: Where does the periodic table end? Discov. Mag. http://discovermagazine.com/2015/march/11-forging-new-elements (2015)

Hoffman, D.C., Ghiorso, A., Seaborg, G.T.: Transuranium People: The Inside Story. Imperial College Press, London (2000)

Hofmann, S.: On Beyond Uranium: Journey to the End of the Periodic Table. Taylor & Francis, London (2002)

Hofmann, S., et al.: New results on elements 111 and 112. Eur. Phys. J. A **14**, 147–157 (2002)

IUPAC: Press release. https://www.iupac.org/cms/wp-content/uploads/2016/01/IUPAC-Press-Release_30Dec2015.pdf (2015)

IUPAC: Press release. https://www.iupac.org/cms/wp-content/uploads/2016/06/Press-Release_Naming-Four-New-Elements_8June2016.pdf (2016)

Jarlskog, C.: Validation of new superheavy elements and IUPAC-IUPAP joint working group. EPJ Web Conf. **131**, 06004 (2016)

Karol, P.J.: The Mendeleev-Seaborg periodic table: Through $Z = 1138$ and beyond. J. Chem. Educ. **79**, 60–63 (2002)

Karol, P.J., et al.: On the claims for discovery of elements 110, 111, 112, 114, 116, and 118. Pure Appl. Chem. **75**, 1601–1611 (2003)

Karol, P.J., et al.: Discovery of the elements with atomic numbers $Z = 113$, 115 and 117. Pure Appl. Chem. **88**, 139–153 (2016)

Koppenol, W.H.: Naming of new elements. Pure Appl. Chem. **74**, 787–791 (2002)

Koppenol, W.H., et al.: How to name new chemical elements. Pure Appl. Chem. **88**, 401–405 (2016)

Kragh, H.: Superheavy elements and the upper limit of the periodic table: early speculations. Eur. Phys. J. H **38**, 411–431 (2013)

Lemaître, G.: The beginning of the world from the point of view of quantum theory. Nature **127**, 706 (1931)

Loss, R.D., Corish, J.: Names and symbols of the elements with atomic numbers 114 and 116. Pure Appl. Chem. **84**, 1669–1672 (2012)

Merrill, P.: Cosmic chemistry. Astr. Soc. Pacific, Leaflet **2**, 25–28 (1933)

Öhrström, L., Reedijk, J.: Names and symbols of the elements with atomic numbers 113, 115, 117 and 118. Pure Appl. Chem. **88**, 1225–1229 (2016)

Oganessian, YuT, et al.: Observation of the decay of ^{292}116. Phys. Rev. C **63**, 011301 (2000)

Oganessian, YuT, et al.: Experiments on the synthesis of element 115 in the reaction Am(48Ca, xn)291–x115. Phys. Rev. C **69**, 021601 (2004)

Oganessian, YuT, et al.: Synthesis of a new element with atomic number $Z = 117$. Phys. Rev. Lett. **104**, 142502 (2010)

Chapter 7
Some Philosophical Issues

Abstract Research in superheavy elements (SHEs) is not only a highly specialized branch of modern science its history also casts light on problems of a more general nature. One of these problems, a classical one in the history of science, is the uneasy relationship between physics and chemistry in transdisciplinary research. Another problem of a philosophical nature relates to the very meaning of the concept of discovery. The transuranic elements are not discovered in nature, but created or manufactured in the laboratory. What does it imply, more precisely, to assign priority to a certain collaboration for having discovered—or created—a new element? Lastly, one may question if all officially recognised superheavy elements exist in the sense ordinarily associated with the term existence. After all, they have very short lifetimes and disappear almost instantly after having been created. Even though nuclides of the heaviest elements have undoubtedly been identified, it does not follow that they can rightfully be classified as chemical elements on par with ordinary elements.

Keywords Superheavy elements · Transfermium working group
Discovery · Ontology · Existence · Creation

Among the general themes that SHE history casts light on is the relationship between chemistry and physics in the modern era. Of the other themes illuminated by the history of SHE research we first deal with the concept of discovery as related to chemical elements. This is followed by considerations regarding the ontological status of short-lived artificially produced elements and their relation to ordinary elements. Parts of the chapter and especially Sect. 7.3 rely on an earlier paper (Kragh 2017).

H. Kragh, *From Transuranic to Superheavy Elements*, SpringerBriefs in History of Science and Technology, https://doi.org/10.1007/978-3-319-75813-8_7

7.1   Between Physics and Chemistry

There is little doubt that SHE research, and especially in regard to the synthesis of the elements, is basically nuclear physics and has been so since the production of the first transuranic elements more than seventy years ago. In the later development SHE experiments and interpretations of data have relied crucially on advanced accelerators and detectors derived from high energy physics, kinds of instrumentation that are foreign to the research tradition of chemistry. What is more, research contributions to SHE science were and still are predominantly published in physics journals (such as *Physical Review Letters* and *European Physics Journal*) and not in chemistry journals.

Nonetheless, SHEs are about elements and there is a long historical tradition that everything concerning new elements belong to the domain of chemistry. Radioactivity is a property of certain elements which spontaneously transmute into other elements and this was the rationale for counting much of the early research in radioactivity to chemistry rather than physics. The physicist Marie Curie received the 1911 Nobel Prize in chemistry, and in 1922 the same prize was awarded to Francis Aston, another physicist. Rutherford did not hold chemistry in high esteem and yet his 1908 Nobel Prize was in chemistry, not physics. "I have dealt with many different transformations with various periods of time," Rutherford reportedly said at the Nobel banquet, "but the quickest that I have met was my own transformation in one moment from a physicist to a chemist" (Jarlskog 2008).

Nuclear chemistry, the successor of early radiochemistry, remained a chemical discipline with crucial significance in SHE research. The responsibility of recognizing new elements belonged to IUPAC or its predecessors, and although the interdisciplinary working groups TWG (Transfermium Working Group) and JWP (Joint Working Party) included physicists, the final decision rested with and still rests with IUPAC. By and large its sister union IUPAP stayed on the sideline. The distinction between physics and chemistry in modern SHE research is in some way artificial as workers in the field rarely consider themselves as either physicists or chemists. Still, the relationship between the two sister sciences and their respective organizations is not irrelevant and has not always been harmonious.

As mentioned in Sects. 5.2 and 5.3, in the early 1990s American SHE researchers including Glenn Seaborg, Albert Ghiorso and Paul Karol complained that the TWG panel was dominated by physicists who did not fully appreciate the methods of nuclear chemistry. While previous disagreements between chemists and physicists were rarely openly discussed, recently the subject has been addressed in a remarkably candid way, if this time from the physics point of view. The Swedish theoretical physicist Cecilia Jarlskog has been a member of the Nobel Prize physics committee and she served from 2011 to 2014 as president of IUPAP. Thus, her voice has some significance. In an address given to a Nobel Symposium of 2016 she declared war against the IUPAC managerial staff which she accused of being incompetent in matters concerning SHEs. As we have seen, a similarly low opinion of IUPAC was earlier expressed by Karol, a nuclear chemist, and also by

Russian SHE scientists. According to Jarlskog (2016), not only was IUPAC responsible for the "failure" of the most recent JWP, it had also stolen the credit from the physicists. And that was not all:

Everyone agrees that the discovery of the new superheavy elements, a process which takes ages, lies in the "camp of physicists". Once discovered, the atomic physicists and chemists can deal with the identification of the electronic structure of these elements. As noted earlier, our physicists deserve and need to be publicly acknowledged for their achievements… [but] the IUPAC managerial staff has gone behind our back and violated the "ethical rules", in spite of not having the competence to validate the discoveries.

Jarlskog consequently suggested that IUPAP and not IUPAC should take the leading role in the SHE enterprise. Only the future can tell whether this will happen or not. When Jarlskog entered the negotiations with IUPAC to establish a new working group, she "could not imagine that there would be 'political' aspects in it." She admitted being "naïve and believing that scientists are impartial and logical." If she really believed it, it was indeed a naïve view.

7.2 The Discovery of an Element

The term "discovery" is used in different ways, sometimes referring to the creative phase of inquiry and sometimes to the outcome of such inquiry (Schickore 2014). Another ambiguity concerns what is discoverable and what is not. Novel objects and phenomena in nature can be discovered, but so can laws and theories. We shall here disregard the concept of scientific discovery when it refers solely to the creative act and to theoretical entities.

To discover some object X is to observe for the first time that X exists and to demonstrate the claimed observation convincingly, meaning to the satisfaction of the relevant scientific community. (I shall here disregard the discovery concept in relation to laws and theories.) It follows that it is not sufficient just to postulate or hypothesize that X exists, as this will not satisfy the community. Moreover, the discoverer or discoverers must not only have observed X but also recognised that X is a novel object or phenomenon with certain properties. The object in question obviously has to exist at the time of the discovery, but not necessarily at an earlier or later time (Sect. 5.2). Or perhaps one should say, more cautiously, that most people at the time of the discovery have to believe that the object exists. The philosopher Peter Achinstein (2001, p. 405) emphasises that what does not exist cannot be discovered. Although that may appear as obvious or even trivial, the claim is not without problems as it presupposes that our current understanding of what exists is permanent. At least from a historical perspective it makes sense to say that chemists discovered phlogiston around 1730 or that physicists around 1913 discovered that the atomic nucleus consisted of protons and electrons. These were discoveries, but wrong discoveries.

Today it is generally accepted that discoveries are not necessarily stable events or assignable to one or more definite discoverers. A novel observation of X made by one person may later be made by another person in ignorance of the previous person's work, in which case we say that X was re-discovered (Olby 1989). In many cases a discovery is made nearly simultaneously by several, more or less independent discoverers, such as there are several cases of in SHE history. There are also examples of what may be called de-discoveries, namely that what was accepted as a discovery was later robbed this status and declared a non-discovery. This is what happened with element 118 in the brief period from 1999 to about 2002. Another example of quite a different kind may be triatomic hydrogen H_3, an unusual molecule which was widely believed to have been discovered in the early 1920s. But a decade later it was de-discovered, only to be rediscovered nearly four decades later (Kragh 2012).

It is useful to follow Thomas Kuhn's distinction between discoveries of objects that were predicted or anticipated and those which were not. SHE discoveries as well as the discovery of sub-uranic elements which filled empty places in the periodic table belong to the first category. The discoveries of argon and helium belong to the first. Unfortunately Kuhn (1962) wrote about discoveries of the first kind that they have rarely been the subjects of priority disputes and that "only a paucity of data can prevent the historian from ascribing them to a particular time and place." As witnessed by the discovery of SHEs and several other elements, this is not the case. Priority disputes seem to be as common for the first kind of discovery as they are for the second kind.

The concept of discovery plays a central role in the reward system of science but is rarely defined or problematized. For example, Nobel prizes in science are explicitly awarded for discoveries and yet the Nobel Foundation has never formulated criteria for discovery similar to those found in the TWG reports of the early 1990s. The authors of these reports were fully aware of the complexity of the discovery concept, which they discussed with no less sophistication than textbooks in philosophy of science. Here is an example (Wilkinson et al. 1993, p. 1759):

> A discovery is not always a single, simply identifiable event or even the culmination of a series of researches in a single institution, but may rather be the product of several series of investigations, perhaps in several institutions, perhaps over several years, that has cumulatively brought the scientific community to the belief that the formation of a new element has indeed been established. However, since different sections of the scientific community may have different views as to the importance and reliability of interpretation of different sorts of scientific evidence, the bringing into that belief of these different sections of the community may well occur at different times and at different stages of the accumulation of the evidence. Where, then, does discovery lie?

It follows that at least in some cases the notion of absolute priority is unrealistic. Those who demand absolute priority to be assigned in all cases cling to "outmoded concepts of the nature of discovery" (Wapstra et al. 1991, p. 883). As we have seen, in several cases the TWG and the JWP chose to assign priority not to a single group but to groups jointly. This extension of the traditional discovery concept had been foreshadowed by the Dubna scientists: "After criteria have been worked out it is quite possible to have the situation where the work of two different groups together satisfies these criteria, while the work of one group does not constitute a discovery. Then it is necessary to consider the concept of a 'joint discovery'" (Flerov et al. 1991, p. 455).

With regard to priority and chronology it often happens that credit for a discovery is assigned retrospectively, in the sense that an earlier non-recognised discovery claim turns into a recognised one because of later work substantiating the original claim. The later work may be due to the discoverer or someone else. We say that James Chadwick discovered the neutron in 1932, but in reality it was only experiments of 1933–1934 which showed that he had detected the particle. The problem with this kind of retrospective discovery credit was admitted by the JWP in one of its reports (Barber et al. 2009, p. 1339):

> The TWG recognized and the JWP strongly continues to agree that there will be situations in which an early paper did not, at the time, convey conviction of discovery, but that later investigations revealed to have been correct. The existence of the element in question is then definitely established by subsequent work following the lead of the early paper. ... The TWG felt it would clearly be wrong to assign absolute priority to that early paper, but that it would be appropriate to weigh its seminal importance.

Whatever the kind of object, a claim does not constitute a discovery before it has won broad recognition, which typically means that it has been independently confirmed by other scientists and accepted by the scientific community. For this reason the discovery claim and the reasons for it have to be announced in public, which normally happens in the form of a scientific paper but can also be in a preprint or in an oral presentation. The inevitable social dimension of the discovery concept can be boiled down to the statement that only those who can convince their scientific peers are credited with having made a discovery. This does not imply that the epistemic dimension is irrelevant. According to a social-constructivist view, "discoveries are ... socially defined and recognized productions ... [and they] occur because they are made to occur socially by processes of social recognition" (Brannigan 1981, p. 77 and p. 169). But although consensus is indeed a necessary element, it goes too far to claim that it is also sufficient and independent of whether the discovery claim is epistemically justified or not.

As realised by the TWG a discovery is not a discrete event which occurs almost instantaneously. It is a cognitive as well as social process that involves recognition from other scientists and their institutions. William Ramsay discovered the new gaseous element helium in the spring of 1895, but even in this relatively uncomplicated case it is difficult to identify a definite moment of discovery (see Kragh 2009 for details). When Ramsay on 22 March identified new spectral lines in a gas evolved from the mineral cleveite, he realised that there was something new in the gas but only after two weeks of further work did he conclude that it was a new element. He then communicated his conclusion or discovery claim in private letters and on 25 April he announced it in a brief paper. The discovery claim was criticized by a few chemists but soon it was confirmed and generally accepted. Ramsay undoubtedly discovered a new element in the spring of 1895.

The discovery of element 72 by George Hevesy and Dirk Coster in early 1923 was more controversial because it was part of a long and bitter priority conflict which involved the very nature of the element (Kragh 1980). Was it a zirconium homologue, hence hafnium, or a rare earth metal, hence celtium? The element was the first one identified by means of its atomic number as revealed by its characteristic X-ray spectrum.

Although by 1925 hafnium was accepted by the large majority of scientists, it still missed the official discovery stamp from the International Committee on Chemical Elements (ICCE), a branch under IUPAC and with several members sympathetic to the rival celtium claimed by French scientists. On the other hand, hafnium was quickly recognised by the German Atomic Weight Commission. In an attempt to secure neutrality, ICCE simply omitted element 72 from its 1925 version of the periodic table! The International Committee chose a compromise in the politically sensitive question, namely by accepting two names and two symbols (Hf, Ct) for the same element. As the committee tersely stated (ICCE 1925, p. 597), "The table does not include No. 72, Hafnium or Celtium, atomic weight 180.8." Only in 1931, at a time when celtium survived in French chemistry only, did a new Committee on Atomic Weights approve hafnium as the only name for element 72. Clearly, there were predecessors to the later discovery and naming disputes concerning SHEs.

The story of SHEs illustrates how the concept of discovery was turned into an operational concept in a particular area of science and how scientists were forced to reflect on the meaning of discovery. The definition of an element did not change as the atomic number Z was still the defining property, but the TWG report of 1991 pointed out that "The exact value of Z need not be determined, only that it is different from all Z-values observed before, beyond reasonable doubt." That determination of the atomic number was still important is shown by the competing claims for having found element 113. When the JWP decided to attribute the discovery to the RIKEN team and not to the Dubna team, it was primarily because the first team provided solid evidence for the atomic number. The Dubna measurements, on the other hand, "were not able to within reasonable doubt determine Z" (Karol et al. 2016, p. 146).

The 1991 report was not only an attempt to define SHE discoveries operationally it also included a discussion of discovery in more general terms (Wapstra et al. 1991; see also Sect. 5.2). The criteria proposed by Wilkinson and his colleagues were in part conventional in so far that they referred to the public domain and to the importance of confirmation. But the report was also unconventional in the sense that it admitted that there were different legitimate views regarding "the accumulation of evidence at which conviction is reached." The evidence had to be solid and reliable, of course, but the interpretation of it could well be disputed.

Confirmation is only meaningful if experimental results are reproducible, but the TWG realised that although reproducibility is an important desideratum it is not always possible or necessary. If a SHE experiment resulted in good data based on unobjectionable methods, confirmation could be dispensed with. This was a situation known from other parts of the physical sciences, such as astrophysics and high energy physics, where complicated and very expensive experiments could not be easily replicated if at all. The complex nature of the discovery concept is further illustrated by the TWG's and the JWP's description of SHE discoveries as cumulative processes. Contributions from one laboratory to measurements obtained by another laboratory might eventually increase the confidence of the results to such a level that they constitute convincing evidence for the discovery of a new element. In such a case, who should be credited as the discoverer? As we saw in connection with the elements from $Z = 103$ to $Z = 105$, the TWG decided to share the credit between two competing teams, one from Dubna and the other from Berkeley and Livermore. The question of who discovered the elements and when cannot be answered with just a name and a date.

7.3   Does Oganesson Exist?

Together with a few sub-uranic elements, the transuranic elements do not exist in nature but are artificially produced in laboratories. They nonetheless count as chemical elements inhabiting their proper places in the periodic table. We say that technetium was *discovered* in 1937 and fermium in 1955, but clearly the elements were discovered in a different sense than gallium and hafnium were discovered. SHEs and artificial elements generally were *created* or *invented*, in largely the same way that a statue is created or a technological device invented. They belong to what the ancient Greeks called *techne* (human-made objects or imitation of nature) and not to *physis* (nature). To Aristotle and his contemporaries, *techne* denoted primarily a kind of craft or skill that could bring forth an artefact from the material nature. While an olive tree was *physis*, a vase was *techne*. Should long-lived SHEs be discovered in nature, unlikely as it is, these hypothetical elements would have been discovered in the traditional sense (or perhaps in the sense that technetium was re-discovered). But the short-lived isotopes below the island of stability would still belong to created and not discovered elements.

Interestingly, Seaborg insisted that the transuranic elements he and others had found were created rather than discovered. "After all," he said, "you can't discover something that doesn't exist in nature any more than Michelangelo discovered his David inside a block of marble" (Johnson 2002). And yet Michelangelo did not think of his famous sculpture as just imposing form onto a lifeless block of marble, but rather as releasing a form that was imprisoned in the block. He reputedly said that he just cut away everything that was not David (Mitcham 1994, p. 127). By contrast, it makes no sense to say that a transuranic element is imprisoned in the nuclear reactants out of which it eventually emerges.

The creation of synthetic and yet in a sense natural objects did not start with the work of Segré and Perrier in 1937, for at that time there already was a long tradition in organic chemistry of synthesising chemical compounds. The first such compound without a counterpart in nature may have been William Perkin's famous discovery (or manufacture) of the aniline dye mauveine in 1856. The discovery initiated the synthetic revolution in chemistry, a revolution which has resulted in millions of man-made molecules. In a sense the synthesis of transuranic elements is a continuation of the tradition in synthetic organic chemistry, only at a more fundamental level.

When Mendeleev and his followers predicted from the periodic table that certain missing elements actually existed, they implicitly relied on a version of the so-called principle of plenitude (Benfey 1965). According to this metaphysical principle as expounded by Leibniz and others, what can possibly exist does exist. Nature abhors un-actualized possibilities. Or, in its modern version, if a hypothetical object is not ruled out by laws of nature it (most likely) will exist and thus be a real object. According to Leonard Susskind (2006, p. 177), a physicist and cosmologist, somewhere in the universe there are planets made of pure gold, for "they are possible objects consistent with the Laws of Physics." In this line of reasoning it is presupposed that existence refers to nature, but the situation with respect to SHEs is different as these elements are possible and yet not realised in nature. The potential existence is turned into actual existence not by finding a SHE in nature, as ordinary elements like gallium and germanium were found, but by creating it in the laboratory. The classical plenitude principle, expressing a belief in nature's richness and continuity, does not seem applicable to the artificial world created by chemists and physicists (Le Poidevin 2005).

Whereas plutonium may be said to be a technological product in so far that technologies are always purposeful and oriented towards social practices, this is not the case with most of the SHEs. They have been produced in minute amounts only and serve no social or economic purposes. More importantly, they have short life-times and thus in several cases have been produced only to disappear again almost instantly. To quote two SHE specialists (Armbruster and Münzenberg 1989, p. 69):

> Once synthesized, elements such as 109 decay so rapidly that synthesis cannot keep up with decay. The heavier elements are so short-lived that by the end of the irradiation all atoms created have already decayed. These atoms must therefore be detected and identified during the production process itself.

Notice that the authors write of "atoms created," whereas in reality what are created are atomic nuclei. They also and consistently describe the created particles as isotopes rather than using the more appropriate term "nuclides." In the 1950s it was widely believed that all transuranic elements existed at the formation of the Earth but that most of them had soon disappeared by radioactive decay. In the press release from the University of California announcing the 1955 discovery of element 101, it was said: "The 17 atoms of the new element all decayed, of course, and the 'new' element is for the present extinct once again" (Maglich 1972, p. 244).

The half-lives of the longest-lived transfermium isotopes vary greatly and generally decrease with the atomic number, from 51.5 days for Md-258 to 0.7 ms for Og-294. The elements have been produced and detected in nuclear processes and thus *did exist* at the time of the detection. But strictly speaking they *do not exist* presently any more than dinosaurs exist. Or perhaps an analogy to the Colossus of

Rhodes, one of the seven wonders of the ancient world, is more appropriate as the enormous statue was man-made and not of natural origin. The Colossus existed for a short period of time (280 BC to 226 BC), but it does not exist. What matters is that the existence of some SHEs is ephemeral or perhaps potential, which is quite different from the existence of ordinary elements whether radioactive or not. Can we truly say that the element oganesson *exists* when there is not, in all likelihood, a single atom of it in the entire universe? Sure, more atoms or rather nuclei of element 118 could be produced by replicating or modifying the Dubna experiments. But within a fraction of a second the re-created oganesson atoms would disappear again.

One may object that particles with even shorter lifetimes are known from high energy physics without physicists doubting that they really exist. For example, the neutral pion decays into two gamma quanta with a lifetime of about 10^{-16} s:

$$\pi^0 \rightarrow \gamma + \gamma$$

The particle was first detected in nuclear reactions in Berkeley in 1950, but contrary to the nuclides of the SHEs it was also found in nature, namely in cosmic rays. The neutral pion thus exists and is not exclusively a laboratory product.

The same is the case with the antiproton $\bar{p}$, another exotic particle first produced in an accelerator experiment and only subsequently detected in the cosmic rays. Incidentally, in 1959 Owen Chamberlin and Emilio Segré—the co-discoverer of the elements technetium and astatine—were awarded the physics Nobel Prize for the antiproton experiment. The antiproton can be brought to combine with a positron and thus form anti-hydrogen according to

$$\bar{p} + e^+ \rightarrow \bar{H}$$

This exotic atomic system has been produced in the laboratory and studied experimentally (Amoretti et al. 2002; Close 2009, pp. 80–100). Anti-hydrogen atoms can under laboratory conditions survive for as long as 20 min. In 2011 an international collaboration of physicists reported observation of 18 events of artificially produced anti-helium nuclei consisting of two antiprotons and two antineutrons, but no anti-helium atom has been detected so far. Anti-hydrogen has in common with SHEs that it is element-like and only exists when manufactured. But contrary to the SHEs, there is no place for anti-hydrogen or other anti-elements in the periodic system.

There is also no place in the periodic system for other exotic atoms where the constituent protons and electrons are replaced by elementary particles such as positrons and muons. Positronium, a bound system of an electron and a positron, was discovered experimentally in 1951 but had been hypothesised almost twenty years earlier (for the early history of positronium, see Kragh 1990). Sometimes described as a very light isotope of ordinary hydrogen or protium, the short-lived positronium has been extensively researched and its chemistry is well known. The positron can be replaced by a positively charged muon in which case one obtains muonium with an atomic mass $A \cong 0.11$ between positronium of $A \cong 0.001$ and protium of $A = 1$. First detected in 1960, muonium has a half-life of about 2×10^{-6} s.

Although muonium does not count as an ordinary chemical element it does have chemical properties and has even been assigned a chemical symbol (Mu). The analogy between muonic atoms and SHEs is underlined by the fact that the nomenclature of the first kind of atoms and their chemical compounds has been considered by IUPAC (Koppenol et al. 2001). In 1970 the distinguished Russian nuclear chemist Vitalii Goldanski wrote a paper on SHEs and exotic atoms in which he suggested that Mendeleev's table remained unaffected by the discovery of the latter kind of atoms. He wrote as follows (Goldanski 1970):

> The replacement of electrons with other negative particles (for example, μ^- or π^- mesons) does not involve a change in the nuclear charge, which determines the position of an element in the periodic system. As to the replacement of a proton with other positive particles, for example, a positron (e^+) or μ^+ meson, such a replacement leads to the formation of atoms which in the chemical sense can be considered as isotopes of hydrogen. … On the basis of the value of the positive charge [positronium and hydrogen] occupy one and the same place in the periodic table.

Goldanski's view is remarkable but also problematic, to say the least. After all, can there be two widely different elements in the same box of the periodic table?

If only a historical curiosity, ideas of exotic chemical elements had much earlier been entertained by a few chemists suggesting that the electron was such an element. This was what Janne Rydberg, the Swedish physicist and chemist, proposed in 1906, assigning the symbol E for the electron and placing it in the same group as oxygen (Rydberg 1906). Two years later, the Nobel Prize laureate William Ramsay (1908, p. 778) independently argued that "Electrons are atoms of the chemical element, electricity; they possess mass; they form compounds with other elements; … the electron may be assigned the symbol 'E'." But nothing came out of these speculations and when the atomic number was introduced in 1913, they were relegated to the graveyard of forgotten chemical ideas.

Another question is whether short-lived transfermium elements really count as *elements* in the traditional meaning of the term. Elements consist of atoms and it is the atoms and their combinations which endow elements with chemical properties. An isolated atomic nucleus has no chemistry. This is what Wilkinson and his TWG stated in its 1991 report and more recently two nuclear chemists elaborated as follows (Türler and Pershina 2013):

> The place an element occupies in the Periodic Table is not only defined by its atomic number, i.e. the number of protons in the nucleus, but also by its electronic configuration, which defines its chemical properties. Strictly speaking, a new element is assigned its proper place only after its chemical properties have been sufficiently investigated.

To phrase the point differently, although a chemical element is defined by its atomic number, not everything with an atomic number is an element.

The point is worth noticing as nuclear scientists commonly refer to an atomic nucleus or a nuclide as were it an element. For example, the 1991 TWG definition stated that a chemical element had been discovered when the existence of a *nuclide* had been identified. However, the term nuclide, coined in 1947, refers to a species of nucleus and thus emphasises nuclear properties. By contrast, the corresponding and older term *isotope* denotes an atomic concept and emphasises chemical properties. The difference between the two terms is more than just a semantic detail, but unfortunately the terms are often used indiscriminately.

Not only is the number of produced transfermium atoms extremely small, what are directly formed are nuclei and not atoms. Under normal circumstances a bare atomic nucleus will attract electrons and form an atom, but the circumstances of SHE experiments are not always normal and the few atoms may only exist for such a small period of time that they cannot be examined experimentally. No atoms are known for the heaviest of the SHEs of which only atomic nuclei have been produced and studied. To this date, some 35 nuclei of element 116, livermorium, all with half-lives less than 50 ms, have been observed.

Despite the elusive nature of SHEs nuclear scientists have succeeded in measuring some of their chemical and physical properties. Ionization potentials have been measured up to lawrencium ($Z = 103$) and even an element as heavy as flerovium ($Z = 114$) has been the object of experimental study. Much is known also about other SHEs, but for a few of them the knowledge is exclusively in the form of theoretical predictions, extrapolations and estimates. For example, in the case of tennessine, element 117, its oxidation states have been predicted to be +1, +3, and +5; the electron structure and radius of the atom have been calculated and so have the boiling point and density of the element as well as of hypothetical compounds such as TsH and TsF_3. But there are no empirical data and none are expected to come in the foreseeable future. Of course, the situation is different for the less heavy transuranic elements and especially for plutonium. Several of the transactinides such as rutherfordium, dubnium, hassium and flerovium have a real chemistry (Kratz 2011; Schädel 2015).

Consider again the heaviest of the elements, oganesson, which is presently known only as one nuclide with an extremely small lifetime. To repeat, very few of the nuclei have been produced and none of them exist any longer. Oganesson has received official recognition from IUPAC and entered the periodic table alongside other and less exotic elements. And yet one may sensibly ask if oganesson is really a chemical element in the ordinary sense of the term. Perhaps its proper status is better characterised as a potential element, something along the line recently suggested by Amihud Gilead (2016). In the spirit of the principle of plenitude Gilead suggests that SHEs exist as "chemical pure possibilities" whether or not they are synthesised and thus turned into actual elements amenable to experiment. Apparently he endows any theoretically predicted atom, whatever its atomic number, with reality. Gilead's concept of "panenmentalist realism" may be of philosophical interest, but it is far from the idea of reality adopted by most chemists and physicists.

In any case, in this essay I am not arguing for an anti-realist position with regard to the SHEs at the end of the periodic table. Nuclides of these elements undoubtedly exist, or rather they existed at the time of their detection, but it is questionable if they exist or existed as proper chemical elements.

References

Achinstein, P.: Who really discovered the electron? In: Buchwald, J., Warwick, A. (eds.) Histories of the Electron: The Birth of Microphysics, 403–424. MIT Press, Cambridge (2001)

Amoretti, M., et al.: Production and detection of cold antihydrogen atoms. Nature **419**, 456–459 (2002)

Armbruster, P., Münzenberg, G.: Creating superheavy elements. Sci. Am. **144**, 66–72 (1989) (May)

Barber, R.C., et al.: Discovery of the element with atomic number 112. Pure Appl. Chem. **81**, 1331–1343 (2009)

Benfey, O.T.: 'The great chain of being' and the periodic table of the elements. J. Chem. Educ. **42**, 39–41 (1965)

Brannigan, A.: The Social Basis of Scientific Discoveries. Cambridge University Press, Cambridge (1981)

Close, F.: Antimatter. Oxford University Press, Oxford (2009)

Flerov, G.N., et al.: History of the transfermium elements $Z = 101, 102, 103$. Sov. J. Part. Nucl. **22**, 453–483 (1991)

Gilead, A.: Eka-elements as chemical pure possibilities. Found. Chem. **18**, 183–194 (2016)

Goldanski, V.I.: The periodic system of D. I. Mendeleev and problems of nuclear chemistry. J. Chem. Educ. **47**, 406–417 (1970)

ICCE: International atomic weights 1925. J. Am. Chem. Soc. **47**, 597–601 (1925)

Jarlskog, C.: Lord Rutherford of Nelson, his 1908 Nobel Prize in chemistry, and why he didn't get a second prize. J. Phys. Conf. Ser. **136**, 012001 (2008)

Jarlskog, C.: Validation of new superheavy elements and IUPAC-IUPAP joint working group. EPJ Web Conf. **131**, 06004 (2016)

Johnson, G.: At Lawrence Berkeley, physicists say a colleague took them for a ride. New York Times, 15 Oct D1 (2002)

Karol, P.J., et al.: Discovery of the elements with atomic numbers $Z = 113, 115$ and 117. Pure Appl. Chem. **88**, 139–153 (2016)

Koppenol, W.H., et al.: Names for muonium and hydrogen atoms and their ions. Pure Appl. Chem. **73**, 377–380 (2001)

Kragh, H.: Anatomy of a priority conflict: The case of element 72. Centaurus **23**, 275–301 (1980)

Kragh, H.: From 'electrum' to positronium. J. Chem. Educ. **67**, 196–197 (1990)

Kragh, H.: The solar element: a reconsideration of helium's early history. An. Sci. **66**, 157–182 (2009)

Kragh, H.: To be or not to be: the early history of H_3 and H_3^+. Phil. Trans. R. Soc. A **370**, 5225–5235 (2012)

Kragh, H.: On the ontology of superheavy elements. Substantia **1**, 7–17 (2017)

Kratz, J.V.: Chemistry of transactinides. In: Vértes, A., et al. (eds.) Handbook of Nuclear Chemistry, pp. 925–1004. Springer, Berlin (2011)

Kuhn, T.: Historical structure of scientific discovery. Science **136**, 760–764 (1962)

Le Poidevin, R.: Missing elements and missing premises: a combinatorial argument for the ontological reduction of chemistry. Brit. J. Philos. Sci. **56**, 117–134 (2005)

Maglich, B. (ed.): Adventures in Experimental Physics, vol. 2. Princeton, World Science Education (1972)

Mitcham, C.: Thinking Through Technology. University of Chicago Press, Chicago (1994)

Olby, R.C.: Rediscovery as an historical concept. In: Visser, R., et al. (eds.) New Trends in the History of Science, pp. 197–208. Ropodi, Amsterdam (1989)

Ramsay, W.: The electron as an element. J. Chem. Soc. **93**, 774–788 (1908)

Rydberg, J.R.: Elektron der erste Grundstoff. Håkon Ohlsson, Lund (1906)

Schädel, M.: Chemistry of the superheavy elements. Phil. Trans. R. Soc. A **373**, 20140191 (2015)

Schickore, J.: Scientific discovery. Stanford Encyclopedia of Philosophy. https://plato.stanford.edu/entries/scientific-discovery/#Dis (2014)

Susskind, L.: The Cosmic Landscape. Little, Brown and Co., New York (2006)

Türler, A., Pershina, V.: Advances in the production and chemistry of the heaviest elements. Chem. Rev. **113**, 1237–1312 (2013)

Wapstra, A.H., et al.: Criteria that must be satisfied for the discovery of a new chemical element to be recognized. Pure Appl. Chem. **63**, 879–886 (1991)

Wilkinson, D.H., et al.: Discovery of the transfermium elements. Pure Appl. Chem. **67**, 1757–1814 (1993)

Name Index

Subject Index

H. Kragh, *From Transuranic to Superheavy Elements*, SpringerBriefs in History of Science and Technology, https://doi.org/10.1007/978-3-319-75813-8

Printed in the United States
By Bookmasters